Technology and the
Human Prospect

Christopher Freeman

Technology and the Human Prospect

Essays in Honour of Christopher Freeman

Edited by
Roy M. MacLeod

Frances Pinter (Publishers)
London and Wolfeboro N.H.

First published in Great Britain in 1986 by
Frances Pinter (Publishers) Limited
25 Floral Street, London WC2E 9DS

Published in the United States of America in 1986 by
Frances Pinter (Publishers) Limited
27 South Main Street
Wolfeboro
NH 03894-2069
USA

British Library Cataloguing in Publication Data

Technology and the human prospect: essays in
honour of Christopher Freeman.
1. Technology—Social aspects
I. MacLeod, R. M. (Roy Malcolm)
II. Freeman, Christopher
306'.46 T14.5
ISBN 0-86187-530-3

Library of Congress Cataloging in Publication Data

Technology and the human prospect.
'Essays in honour of Christopher Freeman'—
Contents: Introduction: Economic analysis and technological change / Norman Clark—'Chips' and 'Trajectories' / Keith Pavitt—Innovation in Materials / George Ray—[etc.]
1. Technology—Social aspects. I. MacLeod, Roy M.
II. Freeman, Christopher.
T14.5.T44184 1986 303.4'3 86-8125
ISBN 0-86187-530-3

Typeset by Joshua Associates Limited, Oxford
Printed by Biddles of Guildford Ltd

Contents

Preface

It was in revenge for the theft of fire from heaven, we are told, that Zeus caused Hephaestus to make a woman out of earth, who by her beauty should bring misery upon the human race. Hermes gave her boldness and cunning, Aphrodite gave her beauty, and the gods called her Pandora—or 'all gifted'. When Epimetheus, brother of Prometheus, took her as wife, he accepted from Zeus a wedding gift of a jar which contained every human ill. Alas, the inquisitive Pandora opened the jar from which all miseries spread over the earth. Only Hope remained in the jar, shut quickly before she, too, escaped.

Time and again, this ancient myth has been used to describe the application of science and technology to the purposes of human contrivance. The 'guilt' of accepting Pandora's box lay with Epimetheus, the teacher. In legend, his personal fate is left unclear, but his dilemma remains with us still. From antiquity, it was given to men to learn, and to let their learning guide their fate. In our day, the moral should not be lost on those who both love beauty and teach the use of fire.

Hope still remains in the jar. And, above all, it is a hopeful vision of the future, asserting the promise of mankind, that abides with those who read the work of Christopher Freeman. In this work we find the vision of a 'critical optimist'. And it is to the man, his vision and that corpus of his scholarship, spanning twenty years, that we gather in these pages to do honour.

In two broad ways, his influence has been profound. In the space of two decades we have seen dramatic changes in our perception of the role and consequences of technical change. We are well accustomed to accepting deep uncertainties in our understanding of those factors which underlie the relationship between economic theory and political possibility. We now see societies, East and West, North and South, in 'crisis', with social structures apparently condemned to discontinuities and dislocations. What, in this context, can be the positive role of science and technology, of human wit and invention? How can the forces of knowledge, leased to privilege, be released to the 'relief of man's estate'? From Adam Smith and Marx, Schumpeter and Bernal, from the fugitive brilliance of Laski and perhaps the luminous foresight of Bacon, the work of Christopher

Freeman has drawn inspiration and dedication. For Chris, it seems, economics follows Keynes's definition: 'a method rather than a doctrine, an apparatus of mind, a technique of thinking, rather than a body of settled conclusions.' To the languge of economics, Christopher Freeman has added the message of freedom. The direction and consequences of science and technology give grounds for concern. But the direction of knowledge is in the hands and hearts of men and women. And it is their capacity for direction which we must study, understand and improve. To understand, we must, in the spirit of the Great Instauration, measure. Then we must demonstrate.

We can do this singly, or together. In Chris's case, we have the benefit of both. Emerson once spoke of an institution as the lengthened shadow of a man. Certainly the institution to which he has been attached, as founder and first Director, owes much to his vision. Under his guidance, the Science Policy Research Unit of the University of Sussex rose to the rank of a world-class organization which has made an important mark upon the field of science policy studies. Over two decades, SPRU has grown, stimulated and sustained through difficult times by Chris and his close colleagues. Few in science policy studies in this country, whether in academic life, government or industry, fail to make pilgrimage at one time or another, to its library and research staff. Through its close connections with international organizations in Europe, North America and the Third World, SPRU now enjoys a reputation *primus inter pares.* Its staff, working in interdisciplinary teams, frequently resembles a Cabinet of all the talents. Much of this success can be attributed to Chris's leadership, and to his intellectual style. Perhaps Jean-Jacques Salomon has captured the essence of this, in speaking of him as 'a master without intellectual snobbery, whose knowledge is as sound as his convictions, who imparts the progress of his own research to his pupils with the modesty of someone still learning, and who guides his students through a leadership that is as much moral as intellectual'. Nowhere more, perhaps, does one find this essence distilled, than in his critique of Robert Heilbroner's *Inquiry into the Human Prospect*, published a decade ago. From that memorable review, few citizens of the sumptuary West can come away content to languish in the 'Luxury of Despair'.

To summarizea living man's work is always hazardous. Looking back, some may find in Chris's academic work, beginning in the late 1950s, an intellectual trajectory that has illuminated much of the field in which he has worked. From the NIESR, where from 1959 to 1966 he surveyed research and innovation in different sectors of industry, and laid the foundations of comparative science and technology indicators at OECD, we follow him through the late 1960s at Sussex—years full of promise but little money for science policy of technology assessment—to the early

1970s, the environmental movement and world-wide concern with our supposed 'limits to growth'. We follow him in the mid 1970s to the problems inherent in forecasting and modelling 'world futures', thence, in the early 1980s, to the challenging implications of what we then called the 'new technology' of microelectronics. Eventually, now, to the contradictions of progress, conjured by the spectre of unemployment and de-industrialization. Today, as Luc Soete shows, Chris has turned to the study of 'long waves', to historical patterns that may provide frameworks for thinking about the future of technological innovation and long-term economic growth, and to the prospect of a 'new technological paradigm'.

In retrospect, the intellectual linkages between these areas become clear, and form part of our shared vision of the role of inventiveness and innovation in shaping human choice. For Chris Freeman, these twenty years have been punctuated by a list of critical studies that have tested conventional wisdom and pointed to wider generalization. Among economists of my acquaintance, few are as sensitive to history, and few as critical of historicism. If grand theory has become unfashionable in economics, it is surely not so in other areas of enquiry, where scholars look beyond the contrary imaginations of our day to the deeply structured properties of the *longue durée*. Technology policy studies need the specific instance, the ideographic comment on the particular and the unique, the sector-by-sector appraisal that lends sentience to common sense. But so, too, the field requires conceptual synthesis, theory and system. And these, quite largely through Chris Freeman's influence and stimulus, show signs, at least, of coming.

During the last twenty years, science policy has been given an active voice. But only slowly have governments recognized the difficulties of fathoming the sources of innovation; slowly we have learned ways of measuring its benefits. Chris once concluded a lecture by reversing the well-known aphorism from Marx's *Theses on Feuerbach*: 'Philosophers have sought to *change* the world; the point, however, is to *understand* it.' Arguably—and Chris would be the first to admit discussion of the point—knowledge must precede change.

If one is required, this mandate provides the organizing principle of the essays in this collection, which recognize just four of the areas of Chris's academic concern. Taking a broad interdisciplinary canvas, Norman Clark places Chris in the context of the professional 'revolution' that has occurred in the economics of innovation and technological change. The international interplay between economic, political and intellectual forces that today informs this 'revolution' is taken up in different ways by Keith Pavitt, George Ray and Richard Nelson. What pitfalls await the 'philosopher' who would be 'king' are elegantly examined by Jean-Jacques Salomon. Charles Falk pinpoints the difficulties confronting indicators in an area Chris has pioneered, while Amilcar Herrera, a close friend of SPRU, looks to the 'present crisis' in technology policy from the perspective of the developing world. In contrasting contexts, Marie Jahoda and Bruce Williams then wrestle with the conceptual response of social scientists to the quintessential problems of technological change and unemployment. Finally, Nathan Rosenberg and Luc Soete build, between them, an intellectual bridge, which carries the recent history of economic analysis from Marx to the 'Schumpeterian Freeman'. Whether all recognize this description of Chris, we leave others to judge—for it in any case draws upon a moving picture of an intellectual odyssey that is still far from journey's end.

We take this opportunity of Chris's official retirement to wish him well. In no sense do we intend this as a final tribute; merely a gathering of some of his many friends from around the world. Some, invited, could not attend, or attend in time. One—Yvan Fabian—died before his contribution could be completed; he is sorely missed. Others will, we trust, find other ways of expressing their thanks to Chris. This is, after all, merely formal retirement. The best is yet to be.

Roy MacLeod
Lewes
Epiphany, 1986

Acknowledgements

This volume owes its existence to the generosity of many friends, and not least to the contributors themselves. For their care and counsel during the long process of preparation. I am greatly indebted to Ms Jackie Fuller and Professor Geoffrey Oldham, Director of SPRU. Together, we shared the experience of renewing old acquaintances, and reviving pleasant memories of Sussex and science policy studies. Both were instrumental in the invitation and encouragement of our authors, and in providing biblio-graphical and personal accounts. All final editorial responsibility has, of course, rested with me. Lord Briggs has kindly opened our tribute, as he inaugurated SPRU, with confidence in the future of an institution which is also a part of his past. To those who offered personal recollections of Chris—particularly Jean-Jacques Salomon—I express my particular appreciation.

For permission to draw upon previously published material in the Intro-duction, we acknowledge the courtesy of Basil Blackwell, Ltd. The index has been compiled by Douglas Matthews, and the text has been copy-edited by Susanna Geddes.

Finally, a special vote of thanks must go to those who, remarkably, among a community dedicated to open enquiry, managed to keep our project a secret for so long: to Linda Gardiner, Glynis Flood and Sue Plume of SPRU, and to Sue Oakeshott and Sandra Kyle of Sydney.

Foreword

Christopher Freeman was the first real Director of the Science Policy Research Unit at the University of Sussex which was founded in 1966. No one more deserves a Festschrift than he.

The Unit, usually known affectionately as 'SPRU', had a pre-history. It was thought of at first primarily in relation to the history and philosophy of science. Sussex was a new university which admitted its first intake of students in 1961. At a time when both the history and philosophy of science were neglected, if not ignored, in most British universities, they seemed to have an acceptable place in Sussex. Moreover, what came to be called the philosophy of the founders of the University emphasized two points—first, the importance of bringing together 'arts', 'sciences' and 'social sciences' in an interdisciplinary teaching programme and, second, the need to attract to the campus of a new and innovating university researchers who were concerned with contemporary problems and issues, including the problems and issues of 'the Third World'.

Much thought was devoted to the idea of a new Science Policy Research Unit between 1961 and 1966. Yet behind the thinking there was, of course, a prior record of experience on the part of those involved in the planning. My own experience at the University of Leeds, where I was Professor of Modern History before going to Sussex as Pro-Vice Chancellor (Planning), had awakened in me an interest in and concern, which has never left me, for 'technology' as much as for 'science': in an industrial city that seemed natural as well as right. While I was in Leeds, I brought Dr Donald Cardwell into the University and introduced courses for history specialists, not only on the history of scientific thought but on the economic, social and political implications of technological change. The University of Sussex, the first of Britain's new universities of the 1960s, offered an unprecedented opportunity for pushing such work further than had ever been done in a British university, since from the start it was determined to introduce 'common courses', linking disciplines, as well as specialist courses. The place and the time were propitious, and although the environment was not industrial I was determined to include among the interdisciplinary schools of the University—themselves an innovation—a School of Applied Sciences.

There had to be a realism, however, in the implementation of strategy. Even during the 1960s, exciting though they were, there were restraints on university policies and resources which were imposed by the University Grants Committee. It needed a struggle to persuade the Committee that there should be a School of Applied Sciences in Brighton, and it needed a vigorous campaign inside the University to introduce an 'arts–science scheme' which should draw in all students, whatever their discipline. Above all, it needed detailed 'logistical planning' to make possible the setting up of a Science Policy Research Unit. What could not be done had to be recognized just as clearly as what could be done. Within this context, the University developed the sensible policy of not introducing a new set of subject courses into the curriculum unless it could provide at least four teaching members of faculty in that field, and it appreciated the necessity of looking to outside funding if it were quickly to create a cluster of policy-orientated research centres and institutes.

It proved difficult in practice to justify the recruitment of a group of four teaching members of faculty to concentrate on the history and philosophy of science: small though such a group would have been, it would have been bigger than that in any other university. It also proved equally difficult to secure large enough core funds to guarantee the long-term future of a Science Policy Research Unit. From the start there was a degree of risk, which was accepted before I knew that I was about to become Vice-Chancellor in 1966. Once I became Vice-Chancellor I knew that it was my responsibility to make the Science Policy Research Unit a success: I was more deeply committed to it than anyone else. It was possible then to reach quick decisions. The creation of a new university would have been difficult without them.

There was, of course, a national and international background to what happened in Sussex. Disussion of 'science policy' in Britain quickened during the 1960s, beginning, perhaps in 1963 and 1964, when Harold Wilson, then in opposition, pressed the case for technological change. The 'anatomy of Britain' seemed wrong. So, too, did the psychology. Much of the discussion was unsophisticated: it was only later that policy studies came to be justified on the grounds that decision-making is a complex process. It was always easier to talk than to act, and for all the talk there was no central responsibility for the long-term development of science and technology in Britain when the Science Policy Research Unit was founded.

It was in 1963 that the University of Sussex gathered together an interesting group of people from government, industry and academic institutions to discuss the national and international implications of research and development, and in later and more private discussions the Sussex initiatives were warmly encouraged by the Department of Scientific and

Industrial Research. The University was fortunate too, in being able to look for help to OECD in Paris: the Organization was a centre of ideas and not just a service station. It was leading the way. Indeed, it was in the year that the University of Sussex was founded that the Secretary-General of OECD appointed a high-level group under the chairmanship of Pierre Piganiol to advise him on the nature and significance of science policy and the part that OECD might play in its development.

Christopher Freeman came to Sussex at my invitation straight from the OECD world in which he acted as a consultant, and from the National Institute of Economic and Social Research in London which I knew well as a Council member. He brought with him as secretary Jackie Fuller, who was working with him as a research assistant, and the two of them together very quickly brought the Science Policy Research Unit into life. They had a long-term vision and a deep commitment to what they were doing.

The Unit was concerned with innovation, and it was itself innovatory, with no blueprint to guide it. This was an advantage. Christopher Freeman made the most of it. So, too, did Geoffrey Oldham who soon joined him—also from OECD—as a Senior Research Fellow. His experience as a geophysicist complemented that of Christopher Freeman as an economist. Between them their compass covered Britain, Europe, Canada and the Third World.

It was an essential part of SPRU, as it was of all research activities at Sussex, that members of the Unit were involved in teaching as well as research. The teaching, it was considered, actually helped the research. Both Christopher himself and all the other members of his core team were teaching and they left a mark on pupils, many of whom were to move from the University into both industry and government. The Unit's contribution to graduate and undergraduate teaching was strengthened by the appointment of one of its early Fellows, Roy MacLeod, to a Readership in the History and Social Studies of Science. For over a decade, from 1970, many SPRU students read for higher degrees through this Subject Group, on subjects in the history, philosophy and sociology of science, as well as in science and technology policy. In 1982, the Unit became responsible for its own graduate programme in science and technology policy studies, and today it mounts a substantial effort in this field.

The Unit grew in size and stature during the nine years of my Vice Chancellorship at Sussex. It was, and is, dependent for its success on short-term project support, backed by limited public funding. It has, however, prospered and, by 1985, its total staff and student numbers had reached one hundred. The University's contribution has now increased to 15 per cent of its total budget.

But statistics are only the bare bones of history. A full history of the

Science Policy Research Unit, complete with 'lessons', would be the kind of interesting case study in which the Unit has specialized. The very first Annual Report published in 1967 is a historic document. 'The primary aim of the Unit', it stated,

> is to contribute through its research to the advancement of knowledge in the sphere of science policy and especially to a deeper understanding of the complex social process of research, invention, development and innovation. It aims to study this process in industry and in government, as well as in universities, and in the context of the environment in developing countries as well as in industrialised societies.

The reference to universities as centres of innovation was characteristic of Sussex—and of the period. So, too, was the stress on 'the joint research of natural and social scientists' and on the necessary links between research and teaching. There was no stress at this first point on forecasting, although it was to become a major research area of the Unit and one which was to secure international recognition. There was, indeed, a continuing historical emphasis. 'Whilst more of its work is focused on contemporary problems of science policy, it is also concerned with the historical evolution of the scientific community, of its professional organisations and of the advisory and executive organs of government concerned with the formulation of science policy.' This essential historical dimension was vital both to the practical work of the Unit and also to Christopher Freeman himself, whose vision of the future encompassed a need to understand the past.

The successive *Annual Reports* of the Unit are full enough to chronicle the story, though not to assess it. They bring out the interdisciplinarity of much of the Unit's problem-centred work. The 1971 Report is particularly interesting. Written in the year that J. D. Bernal died, it stated firmly and explicitly for the first time that the work of the Unit was 'problem-oriented rather than discipline-oriented'. It noted that members of the Unit both from the natural sciences and the social sciences had collaborated with colleagues in Edinburgh to establish two new journals, *Science Studies* and *Research Policy*. There is further interest in this report in that it described the setting up of a social and technological forecasting group, backed by funds from the Social Science Research Council and the Leverhulme Trust. The group was to consider 'alternatives for the future' and the need to develop methodologies 'both for the assessment of available choices and examination of political and institutional arrangements for implementing choice'.

In 1976, the year I left Sussex to return to Oxford, a ten-year review appeared. It is already dated, but it sorted out some of the problems that the Unit had faced in its own brief history and offered a 'critique' as well as presented the facts. The principal research areas were by then clearly identified—'social and technological alternatives for the future' (including 'forecasting techniques', along with 'future patterns of social organisation', 'socio-technical sectors' and 'technology assessment'); 'science and technology policy in developing countries'; 'industrial innovation studies'; and, last in the list, the first area which had been scanned before 1966—'historical studies of science and science policy'. The Rothschild Report (A Framework for Government Research and Development, 1971) had recently offered official encouragement for such a pattern of research, but Christopher Freeman still found it necessary to point out that governments were often reluctant to sponsor independent critical work and that universities might be by-passed or ignored as centres for policy-related research. 'There is scarcely any area of public policy', he declared—and it was one of the firmest statements that he ever made—'where the basic understanding of the issues could not be substantially improved by good research, and there are many where even an elementary theoretical understanding is almost completely lacking'.

It was a great encouragement to the Science Policy Research Unit that in the year when this review was written an independent review panel, chaired by Sir Brian (later Lord) Flowers, Rector of Imperial College of Science and Technology, London, praised the work of the Unit, recognized its international reputation, and encouraged a few other universities to move into this field 'in depth'. At the same time, it observed rightly that since the Unit had had to rely heavily on short-term external support it had lacked the time and resources 'to complete the desirable analysis and synthesis of all the work undertaken'. In asking the Unit to put forward an outline programme of work for the next ten years it stated unequivocally that 'we have no hesitation in recommending to the University that the Unit should continue in being'. In language beginning to be used more often, it was a proven centre of excellence, although 'its modest size, and consequently its coherence' was 'one of its most valuable characteristics'.

I will always associate the word 'modest' as well as the word 'visionary' with Christopher Freeman. What he accomplished at Sussex was always accomplished without fuss. There was no seeking after glory. Yet the word 'modest' is inadequate to convey any full sense of the publication *Thinking About the Future: A Critique of 'The Limits to Growth'* (1973) which perhaps more than any single publication of the Unit established the international reputation of the Unit. It spoke for itself, but it spoke loud and clear. As I wrote in the introduction to it, the thirteen essayists represented in this volume were critical of *The Limits to Growth*

> not because they wish to score points but because they wish to clarify complex issues. They are concerned both with assumptions and with

> methodologies. They do not wish the debate in which they are engaged to parallel some of the noisy debates of the nineteenth century . . . when more heat was generated than light.

There had already been constructive discussions with Dennis and Donella Meadows. They were the kind of discussions that can contribute most to the elucidation of science policy, not least on the biggest and on the most controversial issues which involve values as well as techniques.

Thinking About the Future was published by the short-lived Sussex University Press. SPRU, like the Institute of Development Studies at Sussex—it, too, widely recognized internationally as a genuinely independent centre of research and ideas—has carried on its work through into a very different period from that between 1965 and 1976. Already, of course, conditions in 1976 were very different from those in 1965 for many reasons, including the energy crisis which SPRU had itself foreseen and studied. Yet continuity has been maintained since 1976 in personnel and in purpose.

There was one private benefactor of the Unit, then in his eighties, who deserves to be remembered in any account of the early years, not least because he was always willing, when prompted, to try to think about the future. R. M. Phillips of Brighton, who had had no education in science or social science himself, provided long-term benefactions to the Unit which guaranteed it a secure home in the newly named Mantell Building in the University. Eroded though they were from the start by unprecedented inflation, his annual grants none the less demonstrated a faith in what the Unit was and might become. My very regular tutorials with R. M. Phillips were often the prelude to action in the Mantell Building, named after another Sussex man who left his mark on his own generation. Christopher Freeman has himself become part of a long and rich Sussex history.

Asa Briggs

Introduction: economic analysis and technological change—a review of some recent developments*

Norman Clark

Introduction[1]

Recent years have witnessed an explosion of writing in what may be loosely termed the 'economics of innovation and technological change'. This essay is designed as an introductory review of some of this literature for students and others who have little background in the subject but who may wish to set it within a wider overall context. A major problem for the newcomer, who is often not an economist, lies in understanding precisely what the research is really all about and why it is held to be so important. A central topic is clearly that of economic growth and its causes (and that is where I shall start) but behind this major question there are a range of subsidiary 'issues' concerning the allocation of resources to research and innovation, which inform the main theme and interact in complex ways.

I shall argue that, far from shedding light on social policy towards science and technology, in reality much of this literature (and the debates which it encapsulates) is doctrinal in nature—economists of an 'orthodox' persuasion attempting to reconcile the practice of technological change to conventional neo-classical theorems while those of a more critical bent attempt to theorize in a more 'radical' manner. On the whole, economics, as currently practised, is probably only of little relevance to the understanding of this complex social process, which is inherently interdisciplinary. However, since the failure of economics is really the failure of a cognitive metaphor, and not that of a scientific theory, and since there is currently no adequate alternative, it is likely that the neo-classical tradition will continue to provide the principal organizing 'heuristic' for the study of innovation and technological change. In reality, therefore, the subject still awaits its 'Keynes', although writers like Freeman, Galbraith, Nelson and Rosenberg have done a great deal to pave the way.[2]

* This essay is a shortened and revised version of Chapter 6 of my book *The Political Economy of Science and Technology*, Oxford, Basil Blackwell, 1985. I am grateful to the publisher for giving me permission to draw upon this material. My thanks are also due to Martin Fransman who provided valuable comments on an earlier draft.

Continuation of orthodoxy

After the end of World War II there was a revival of interest in economic growth. A. K. Sen[3] has pointed out that this was not only a reaction to the pre-war debates on short-run economic stability (which Keynes appeared to have won). It was as much, if not more, a symptom of the times. There was the important question of reconstruction of the war-damaged economies of Europe, assisted by American aid through the Marshall Plan. With the growing prospects for decolonization of many parts of the Third World, attention became increasingly focused upon how economic growth took place and upon how the process might be accelerated through judicious action, both national and international.

But progress was rudimentary. So far as the scientific community was concerned, the war experience had brought its members into 'government' to a much greater (and more varied) extent than ever before. However, its involvement was concerned very much with the 'machinery' of warfare, the devices, systems and techniques which had rendered the conduct of hostilities quite different from anything known previously—like radar, underwater systems and, particularly, atomic weapons. Such a community, therefore, was not particularly well-equipped to divert its (latent) talents to the often less mechanistic requirements of post-war reconstruction, and indeed Britain and the United States in particular have continued to devote a very large proportion of their respective expenditures on science and technology to the artefacts of modern war.[4] Nor were the economists particularly progressive. Following Roy Harrod's well-known attempt to chart the conditions under which an industrialized economic system could grow in a stable manner,[5] the economics literature grew replete with increasingly abstruse variations on a theme which had little or no empirical content, and therefore policy relevance.

The economics of Research and Development (R & D)

The break came with the identification of the 'residual' on the part of Solow,[6] Abramovitz[7] and others in the mid-fifties and hence on the need to understand what it consisted of. Clearly R & D was seen to be very relevant. After all, if economic growth is largely *caused* by 'better ways of making things' then investment in this must surely pay off and indeed, since industrial R & D expenditures had been increasing rapidly, it was intuitively plausible to suggest that this must be one of the major factors involved. There followed a series of attempts to establish such an association with, however, very inconclusive results. The evidence tended to show first that what statistical relationship there was between R & D expenditures and economic performance tended to be small compared to other causal factors *and*, more importantly, that causation appeared to work in reverse—namely, improvements in economic performance *leading to* greater expenditures on R & D rather than the other way about.[8] At the 'macro' level, therefore, it almost appeared that R & D represented an item of 'luxury consumption' expenditure on the part of firms.

Certainly, it is not difficult to explain this apparent contradiction, but nevertheless, the question arose that if for a variety of reasons the social productivity of R & D expenditures appeared to be low, why did firms spend money on it? The answer to this appeared to lie in two directions. Firstly, it came to be understood that a good part of *observed* R & D expenditures represented necessary inputs in the productive process rather than expenditures upon innovation *per se*. For example, in industries connected with electronics or electrical equipment one cannot manufacture devices without at the same time possessing facilities for routine trouble-shooting, quality control, testing, calibration and so on. More generally, much of observed R & D expenditures is largely determined by the nature of the industry under consideration.[9] The second explanation was that R & D expenditures represented a necessary element in a firm's competitive strategy either as part of product differentiation activities or as a device to reduce risk in an uncertain environment. Freeman,[10] for example, has argued that in any industry the establishment of a technical lead is a powerful instrument of competitive strength, and we shall see below that Nelson and Winter have recently developed a theory of firm behaviour in which competitive R & D plays an important role.

Certainly these explanations were consistent with the empirical evidence at the 'micro' level where a number of studies, like those associated with Mansfield[11] and Minasian[12], for example, showed quite clear relationships between R & D expenditures and subsequent economic performance. Hence, the general conclusion appeared to be consistent with the proposition that while the *social* productivity of R & D might be small (or even negative), nevertheless, within any one industry or sector it might be dangerous for an enterprise to opt out of R & D spending since that would be tantamount to reducing its *private* productivity. R & D has the properties of an escalator moving downwards at an increasing rate through time. The firm may have to climb it since otherwise it goes out of business, but the effort required to actually move up becomes progressively greater.

But does this conclusion mean that possibilities for economic progress by means of technological change are inherently limited? Certainly, there is an important current of modern thought which has begun seriously to question the whole social and economic role of science and technology;[13] and this relates not only to the so-called 'costs' of technological progress but to a more fundamental critique of a Western civilization which has become so thoroughly impregnated with a 'scientistic' world view that it is unable any more to stand back from itself and evaluate its own development in a disinterested way. 'Science', according to this view, has become an institutionalized and professionalized means of acquiring and legitimizing wealth and power. It has become allied to large bureaucracies, attempting to use its intellectual trappings, including certification, to persuade the uninitiated public of the social value of its services—often at great cost even when apparently successful in so doing. In cases where it attempts to open up its decision-making procedures to a wider public scrutiny, through mechanisms of evaluation such as those developed recently by Irvine and Martin for example,[14] this is seen merely as a means of making the scientific 'machine' more efficient. Deeper questions of the social goals of science and technology remain untouched.

The neo-classical tradition in economics has not begun to relate to issues such as these. Nor would it consider them relevant, if only because the existing intellectual division of labour decrees that they are the concern of others. Conversely, in practice economic orthodoxy has recently taken a much more positive position, arguing that although it is difficult to establish causal links statistically between innovation activity and economic growth, nevertheless there is indirect evidence that such links do exist. It would argue also that innovation is a continuous process and thus not liable to be observed easily at any point of time. For more recent years a relevant argument might be that the current pessimism is the result of recessionary social conditions and tends therefore to overlook innovations which hold out promise for radical (and beneficial) social change, including shortening of the working week. Not to pursue them vigorously would be reactionary in a fundamental sense.

Understanding the residual

However, this position has not been reached without some debate. An early point of concern was a series of attempts to 'explain' the results of Solow and others regarding the (high) relative importance of technological change as an 'explicator' of economic growth. A major purpose of these investigations may have been to reassimilate Solow's conclusions into the mainstream of orthodoxy, although researchers like Denison[15] clearly also felt that if the main components of the residual could be established accurately then this would have important implications for subsequent policy. From a doctrinal viewpoint, however, the problem was that Solow's conclusions diminished the importance of investment (and hence capital) in the growth process, thus creating yet one more 'empty box' for economic theory to puzzle over.

In practice the attack on Solow took the form of criticisms firstly of his functional specification (Cobb–Douglas), and secondly of the way in which he measured his input variables. On the first point it is now widely accepted that there were no systematic biases resulting. However, on the second point there has been more controversy. Measuring the inputs of capital and labour at historic cost, for example, tends to ignore quality improvements resulting from such influences as a better-educated labour-force and on-the-job learning,[16] while authorities such as Stigler[17] have emphasized the difficulties involved in separating off the productivity effects arising as a result of changing relative factor prices.

Part of the problem is clearly semantic (or definitional) since if we define technological progress in terms *only* of 'new technology' then a high residual (called something different, like 'productivity increase') is obviously unacceptable. But to criticize the models of Solow and Abramovitz, on the grounds that sequential additions to the capital stock actually 'embody' new technology, surely smacks of scholasticism. What is important is that economic performance at the aggregate level could not be 'explained' on the basis of conventional variables, clearly suggesting an 'agenda' for research.

Perhaps the most determined effort in this direction was mounted by Denison who in a series of statistical studies has attempted to disaggregate the residual into its constituent parts, a truly heroic exercise in reductionism. One interesting feature of Denison's research has been his 'low' finding for organized R & D as a component of the residual. For example, one study[18] showed that for the American economy between 1929 and 1957 which grew by 2.93 per cent, as much as 2.00 per cent was made up of 0.43 per cent due to increases in capital stock, the remaining 1.57 per cent being due a range of influences of which 'advances in knowledge' (0.58 per cent) and 'economies of scale' (0.43 per cent) were the main ones. Only 0.12 per cent could be ascribed to organized R & D expenditures. Denison's results have been supplemented by more disaggregated studies, like that of Griliches[19] on hybrid corn, where Griliches estimated that over the period 1910–55 'the social rate of return on private and public resources committed to research on this *highly* successful innovation was at least 700%'.[20] More recently, Rosenberg has pointed out that economic historians have conducted a wide range of historical studies into the role of technological changes in a number of industries, although the conclusions are indeterminate.

However, there are two major (and related) problems with Denison's approach. Firstly, he tends to *ascribe* percentage weights to his variables without sufficient discussion given to the methodological basis for prescription. Secondly, it is not clear why the various causative influences can be viewed as acting independently of each other. If they can, then their estimation would provide an invaluable set of guidelines for growth policies. However, if they cannot (and this seems much more likely), then it is not easy to see what value the empirical work has. Indeed, it is difficult to avoid the conclusion that very often results of the kind reached by Denison beg much more fundamental questions about the nature of the methodologies used. It may not be so much a question of Rosenberg's 'Black Boxes' being unfilled, as one of the wrong sorts of boxes being examined altogether.[21]

The role of demand in technological changes

A very important strand of thinking within neo-classical orthodoxy is represented in the work of those who argue that inventions and innovations are 'called forth' by consumer demands as and when required. In this context Rosenberg[22] has pointed out that by consigning the development of science and technology to a 'limbo' of autonomous activity producing a 'shelf' of technologies which may be drawn upon at will, it has thereby become possible to resurrect the primacy of neo-classical economics in the analysis of this rather difficult area of social life.

The person usually given the credit for systematically developing the theory of demand-led technical change is Jacob Schmookler, whose book *Invention and Economic Growth* (Harvard University Press) was published in 1966. Schmookler's aim was to demonstrate, using data on patents and investments for a number of industrial sectors, that variations in consumer demands, as reflected in rates of investment, varied systematically with subsequent variations in patenting activity. This was true both across industrial sectors at any point in time as well as for any one industrial sector through time. His studies revealed high correlation coefficients and hence he concluded that inventive activity could be regarded as similar to any other economic magnitude and hence as being subject to the interplay of market forces:

> . . . the belief that invention, or the production of technology generally, is in most instances, essentially a non-economic activity is false. . . . [The] production of inventions, and much other technological knowledge whether routinized or not . . . is in most instances as much an economic activity as is the production of bread.[23]

A similar sort of argument had previously been put forward by Hessen in 1931[24] but in that case it concerned the pattern of expenditures upon basic science which Hessen argued was determined largely by industrial needs.

Both Schmookler's and Hessen's arguments have been subjected to considerable criticism in recent years,[25] and it is in fact becoming clear that although demand factors do have a role, innovations are not equally available at equivalent cost for all industries. On the contrary, the ease with which new technologies can be assimilated into economic production also depends fundamentally upon the state of development of a variety of scientific and technological sub-disciplines and that varies very widely indeed. By way of caricature, Rosenberg drives home the point in the following way.

> It is unlikely that any amount of money devoted to inventive activity in 1800 could have produced modern, wide-spectrum antibiotics, any more than vast sums of money at that time could have produced a satellite capable of orbiting the moon. The supply of certain classes of inventions is, at some times, completely inelastic—zero output at all levels of prices. Admittedly, extreme cases readily suggest arguments of a *reductio ad absurdum* sort. On the other hand, the purely demand-oriented approach virtually assumes the problem away. The interesting economic situations surely lie in that vast intermediate region of possibilities where supply elasticities are greater than zero but less than infinity![26]

International effects—the product cycle

Controversy regarding the impact of technology was not, however, confined to the domestic economy only but also became involved with international trade and investment. Here it is useful to introduce the notion of the 'product cycle', first put forward as an analytical idea in relation to international trade by Michael Posner[27] in 1961, and then later refined and adapted by Hirsch,[28] Hufbauer,[29] Vernon[30] and others. A major reason for its introduction was to attempt to deal with two major weaknesses in international trade theory as this had developed within the neo-classical tradition: firstly, neo-classical theory did not appear to be a very good 'predictor' of the pattern of trade between countries, and secondly, it had little to say about the dynamics of trade, about why and how trade patterns changed.

Actually, 'technology' appeared to have very little *at all* to do with trade. Trade took place, according to the neo-classicals, because factor endowments varied between countries. A country which was particularly favoured by factor of production X_1, for example, would by that token possess a *comparative advantage* in the production of those commodities for which this was an important quantitative input, since X_1 was *comparatively* cheaper than other factors X_2, X_3, X_4 . . ., and so on. It would then pay that country to specialize in the production of such commodities, exporting the surplus beyond domestic requirements and using the proceeds to import commodities manufactured using those inputs which were comparatively cheaper in other countries. In this way the promotion of international trade would ensure that world resources were optimized and output maximized. It would also tend to equalize the income of any given factor of production across countries. Much of international trade theory was, and still is, a series of articulations and elaborations of this basic theme. Since 'technology' was simply defined in terms of a combination of inputs[31] and was assumed to be freely available at zero cost from an international 'shelf', it played no part by definition in the pattern of trade.

However, as a scientific 'theory', the neo-classical view was unsatisfactory. Leontief,[32] for example, had demonstrated empirically that the United States, a country manifestly well endowed with capital, appeared paradoxically to be a net exporter of labour-intensive commodities and a net importer of capital-intensive commodities. The incomes of factors of production (for example wages) demonstrated a stubborn unwillingness to equalize across international boundaries and it was becoming abundantly evident to all but the most committed of the faithful that 'technology' was by no means freely available at zero cost to all potential producers of a good. Moreover, the established theory provided no account of changing trade patterns through time. It was with this background that the product cycle appeared on the scene.

The argument ran as follows: products do not suddenly appear on markets as fully-fledged mature commodities but rather demonstrate a 'life cycle' of techno-economic maturation which may be approximated as three broad phases of development. In the first phase, the *innovative* phase, the product in question is first placed upon the market as a 'new good'. This tends to happen in advanced economies since only there do you have the technological infrastructure necessary to support product innovation—access to skills, a technically advanced capital goods sector, an abundant supply of risk capital, etc. The advanced economy firm, then, has a comparative advantage in the marketing of new products which it exploits to gain monopolistic rents à la Schumpeter.[33] Prices are high but there also tends to be very rapid growth of output and sales. Since the market is a high-income one, there tends to be buoyant receptivity on the part of consumers.

As time T_1 approaches in Figure A, the monopolistic position of the inventor begins to be eroded as patents run out and other competing firms begin to copy or introduce substitutes. The rate of profit falls along with price, and production moves out of the small, advanced, innovative firm into the larger, integrated, science-based enterprise which has the capacity to benefit from scale economies. This then is Phase II, the *intermediate* phase, when output steadies down to a 'normal' growth rate. The role of science now becomes the more standard one of routine testing, quality control and marginal (defensive) improvements which occupy most of the time of the conventional R & D laboratory. At this stage too there will be some internationalization of production.

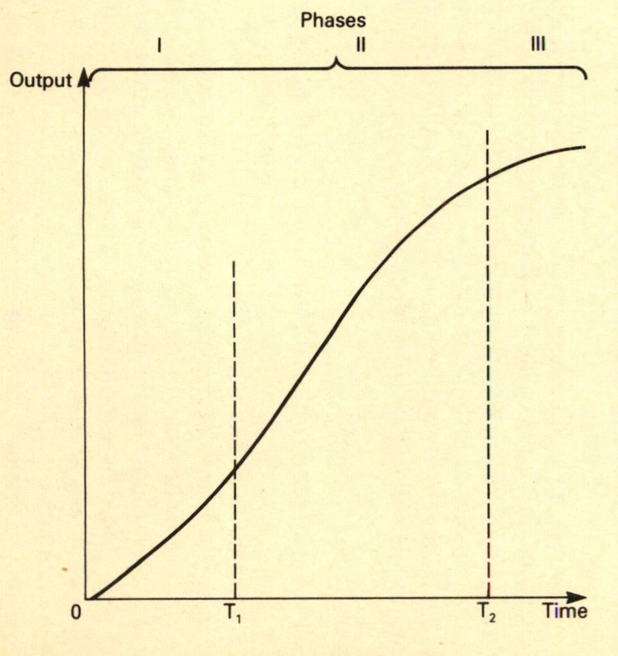

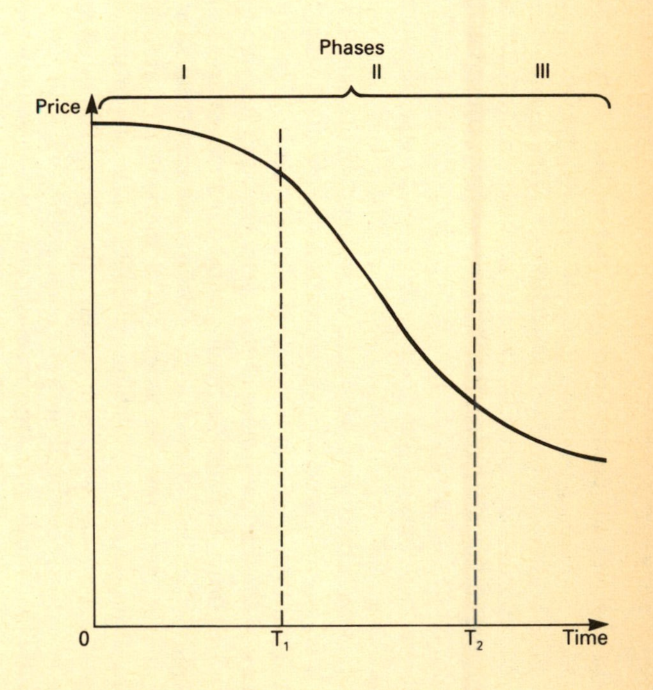

Figure A Phases of product development

Finally, in Phase III, the product becomes fully *mature* in the sense that its production technology is now completely understood and standardized. Possibilities for further innovation are rare, monopolies are eroded, output falls off and price falls to a minimum 'competitive' level. It is at this stage, so the story goes, that underdeveloped countries (LDCs) have a comparative advantage in production since unskilled and semi-skilled labour have become the major inputs, and these of course are cheaper in LDCs.

But product cycle theory was not only an explanation of changing trade patterns. It was also implicitly (and sometimes explicitly) a theoretical statement about the transfer of technology, since this would be expected to take place at the 'mature' phase of the cycle. Cooper[34] has pointed out that in practice firms 'transfer' technology and other resources at a much earlier stage in the 'life' of products, often for reasons connected with defensive investment strategies where local markets are protected. In other cases firms separate off labour-intensive parts of the production process and carry on what are essentially cheap-labour assembly operations overseas, re-exporting the finished parts back to headquarters. Enclave activities of this type are especially common in some far-eastern countries (Taiwan, South Korea, Singapore) and often relate to products which are by no means exhausted from an innovation point of view. A final point is that where certain spheres of economic production are monopolistically controlled by large international firms, technology may not be transferred even where the relevant products are mature in this sense.

For reasons such as these, therefore, product cycle theories are probably only marginally useful in the context of the transfer of technology from rich to poor countries. More modern approaches tend to be more directly policy-focused, emphasizing the ways in which poor countries are able (or not) to develop their own indigenous technological capabilities, and through this enhance their prospects for long-run development.[35]

The diffusion of technology

In recent years the literature on the diffusion of innovations has expanded enormously, and it is worth mentioning briefly one or two of its features at this stage since it relates to much of the current debate about the social impact of science and technology. A useful starting-point is the *uncertainty* which attaches to all innovative activity whether it be the launching of a new product for the first time or the imitation of an innovation already introduced by a competitor. You may recall that Schumpeter placed a lot of emphasis on the competitive pressures which act so as to stimulate the imitative diffusion of an innovation throughout the economic system, but Schumpeter was also very much aware of the risks involved for the firm. Not to adopt a crucial innovation might mean going out of business altogether as existing product types become progressively more obsolete. On the other hand, the act of adoption itself is fraught with its own uncertainties caused by factors like the lack of access to important elements of ancillary technology, unpredictability of demand, possible changed requirements for marketing and distribution, *and*, very importantly, the possibility that further radical breakthroughs may yet be made. Rosenberg[36] has pointed out that under expectations of further radical technological changes it will make sense for a firm *not* to adopt an innovation even where all other risks are low and such action makes clear economic sense.

He cites a series of examples which support the view that an initial 'innovation' is often a very imperfect entity, full of 'bugs' which need to be eradicated and teething problems that require to be solved. At this stage, which has clear similarities with Phase I of the product cycle, the sensible course for potential competitors is to await developments while keeping at the same time a weather eye upon events. Only when the innovation has become relatively 'mature' will it make sense to undertake direct imitative investment. This is often true in practice even for the relatively large firms which, according to writers like Schumpeter,[37] Galbraith[38] and Scherer[39] are better placed to 'hedge' against the risks of innovative failure because of sheer financial power and market control.

Further support for this (rather conservative) view on the rate of diffusion of innovations is given by Freeman who draws attention to the relatively routine and defensive nature of much R & D activity in modern industry. This is in general 'not devoted to the major new product or process but to relatively minor changes and modifications'.[40] Rosenberg[41] himself has documented a whole series of historical cases where the older (and apparently superseded) technology has continued to remain in existence for decades after the new technology came on to the scene. In such cases the rearguard success of the older vintage was almost always due to effective technological improvements on the part of firms which for one reason or another were not able to adopt the new technology.

On the whole, with a small number of exceptions, the neo-classical tradition of economic analysis appears to have had great difficulty engaging with factors such as these, preferring instead to 'model' the diffusion process using established 'economic' variables like profits, sales, investment, etc., mediated by the circumstances of individual firms.[42] Uncertainty is dealt with by the device of 'rational expectations' and most models are usually able to generate the classic 'S'-shaped diffusion curve depicted as something akin to the product cycle function outlined above. A number of critics now wonder, however, whether such activities need to be developed much further, since by eschewing, relatively speaking, how firms actually *do* behave this tradition is in danger of disenfranchising itself from policy relevance. Decision-makers, whether at corporate or government level, need above all to be informed by theories which are real and not simply elegant abstractions.

More radical developments

Readers will have gathered by now that the essence of the 'orthodox' view regarding the impact of science and technology on socio-economic development is that more needs to be done *within the existing paradigm*—i.e. with the methodological and conceptual tools already available. These tend to emphasize the 'comparative static' nature of much of economic theorizing, the relative importance of demand factors which 'call forth' responses on the part of the productive sector, the accompanying passivity and 'shelf-like' nature of technological developments, and the primacy of conventional economic aggregates. They also tend to gloss over important issues of political economy through which technological changes may be viewed as an intrinsic part of the bureaucratic and political operations of the modern industrial economy. There are, however, so many evident problems inherent in this position that a variety of quite radically new approaches have been put forward in recent years. The remainder of this essay is devoted to an outline of the more important of these.

Science push

Diametrically opposite to the notion of the 'technological shelf' is the view that all innovations stem mainly from advances in scientific research. Of course, in its most extreme sense—namely that basic science *causes* technical change—the notion is clearly untenable. Not only does disinterested research produce much work which could by no stretch of the imagination be translated into commercial devices but history is replete with examples of important innovations which manifestly did not emanate from basic scientific research, even if later scientific work may have played a role in bringing the ideas to commercial production. However, the notion of 'science push' is a rather more sophisticated one in which the role of professional R & D is given a primary place. Thus, in a series of detailed industrial case studies, Freeman[43] has shown how in this respect the twentieth century has differed from the nineteenth century, precisely by virtue of a full-time professional R & D sector, attached to industry, whose function it is to search for and articulate commercial innovations by means of organized scientific research.

The starting-point was the very rapid development of new process technology in the chemicals sector towards the end of the nineteenth century.

Beginning with the exploitation of synthetic aniline dyes in the 1860s based on coal-tar chemistry, and then progressing through the synthesis of nitrogenous fertilizers and the catalytic cracking of petroleum, Freeman demonstrates how firms (mainly in Germany and the United States) very quickly came to understand that the 'organized search for knowledge' represented a systematic means of ensuring a continuous flow of innovations, thereby helping to ensure market strength and corporate growth and survival.

Nowadays the importance of design engineering and contracting have led to a very high degree of specialized division of labour as exemplified, for example, in the design and development of the modern nuclear reactor. Much of modern industrial production depends upon the search for, the verification of and the processing of scientific information, and its subsequent articulation in engineering forms. Freeman argues that in areas of economic production which have experienced fast industrial growth, it is the close and pervasive link with a scientific base which really provides the thrust for technological change. The corollary is that governments wishing to promote international industrial competitiveness cannot sit back and await 'entrepreneurs' to do the job by themselves, since the complexity of the innovative act is now such that often only an organized team effort is likely to be successful.

Here he takes issue with writers like the historian Derek de Solla Price who argued in 1965[44] that 'science' and 'technology' have always been two entirely separate professional activities, that contact between them, while occasionally very fruitful, has tended to be spasmodic, and that for long periods technology has progressed without any major inputs from science. He takes issue also with a seminal piece of empirical work carried out by Jewkes *et al.* in 1958[45] who in a study of around sixty important twentieth century innovations concluded that the role of the corporate R & D laboratory had been greatly exaggerated.

However, both writers, according to Freeman, do not take sufficient account of how things have changed as this century has progressed. Price tends to draw his examples from earlier epochs, particularly during the nineteenth century, while Jewkes' own sample is biased in a similar direction, including for example the zip fastener but excluding nuclear technology. Moreover, there are important methodological problems in Jewkes' study concerning the economic importance of the innovations (and therefore their relative 'weighting') and on how much stress should be given to the very important 'development' phase during which the raw invention is turned into a useful commercial product. Freeman cites work by Mansfield[46] (1971) and Stead[47] (1974) which indicate that such R & D costs vary 'typically between 25% and 60% of total innovation costs'[48] so that even in cases where the original 'invention' came from one person, it has become increasingly necessary for substantial 'developmental' resources to be put in before it reaches commercial viability. This tends to take place with the R & D laboratory. Of course, all disputants agree that such trends vary across industry and in particular that the more traditional industries (like textiles) are by their very nature less research-intensive. However, the position of authorities like Freeman is that it is precisely the fast-growing industries like chemicals, pharmaceuticals, plastics, electronics, computers, aircraft and nuclear power generation which are characterized by high in-house R & D spending and which are at the same time linked to advances in basic scientific research. In industries like these the lone inventor is conspicuous by his/her absence.

To the extent that Freeman's view holds on the importance of professional R & D, it offers considerable support to the position of Galbraith and others that large firms are likely to be better innovators than small firms. On the whole the modern evidence appears to support this view although there are areas of high technology where small firms still make substantial inroads, like branches of electronics capital goods. A related point is that industry cannot normally substitute access to publicly-funded scientific research (through universities and other research institutes) for its own in-house efforts. Although there is still a lot of ambiguity regarding the precise nature of the links between university-based research and technological development in firms, it is generally agreed that in-house R & D provides a necessary 'filter' for external R & D inputs. It is only through the act of conducting R & D itself that the firm can keep abreast of progress in the relevant scientific disciplines, recognize the significance of particular 'advances', and have access to the appropriate 'invisible college' of university-based scientists and engineers.[49]

Hence, the notion of 'science push' in its modern form is intended not so much to convey an impression of scientific research *forcing the pace* of industrial innovation, but rather attempts to bring out the much more complex nature of industrial innovations which *relate* to basic science (as most nowadays appear to do). In many sectors firms *must* invest a given absolute sum in R & D, the precise amount determined by the nature of the industry in question, simply because if they do not they will find it increasingly difficult to compete with their rivals. Investment in science does not so much 'push innovation' as it acts as a necessary condition for industrial survival.

Equally, however, the very fact of industrial R & D makes a nonsense of the ideal of the 'technological shelf', which is readily available to all firms, of all sizes and wherever they may be located. A more accurate analogy is that there are many technological shelves whose stocks are constantly being changed as a result of the direct investment action of firms and of the (uncontrollable) actions of others. Sometimes the contents of the shelves are private and appropriable by the firm. Sometimes they are not. Often the existence of the shelves and their contents are simply not known about. In reality, therefore, all firms live in a world of great technological uncertainty. How to cope with this uncertainty is thus a major problem.

Technological systems

A notion which is related to that of science push, although conceptually distinct from it, is that of the technological 'system', sometimes also called the technological 'trajectory' or 'family'. This concept is intended to convey the idea of a network of proven engineering relationships which have been articulated into a variety of designs, blueprints, systems and sub-systems of machinery, jigs and fixtures, and agendas for engineering research, all held together by a common technological thread which has shown some promise in economic production. Examples of a technological system might be machine tool technology in the nineteenth century, transistor technology in the 1950s and 1960s, and computer aided design technology in recent years. In each case the technology cannot be regarded as an invariant 'piece of know-how' which is applied to economic production in a mechanistic fashion, but instead shows all the hallmarks of an organic system, often associated with a new industry, but not necessarily, which develops along paths determined by and unique to itself often in a very speculative 'trial and error' manner. As Rosenberg has written, the history of technology shows clearly that this is the case in most areas of economic production.[50]

On this view the role of the market is to act as a series of 'guiding posts' or 'filters' which will constrain the technological trajectory as it is being developed to those innovative activities which look likely to be successful.[51] But 'success' need not be immediate, of course. This is why the actual decision-making on R & D in real life often takes on the characteristics of a political battle (but within the firm), with different factions arguing the respective merits of their own particular projects, and drawing upon whatever evidential and institutional support they can muster so as to win the day. Often, firms handle the inherent uncertainty involved in R & D investment by establishing a number of concurrent research projects. As each of these develops, some begin to appear more promising than the rest so that gradually resources are concentrated upon the 'winners' while the 'losers' are brought to a halt.[52]

Nor is the market the only feature of such a selective environment. In cases where the 'productive act' is not subject to market forces (for example weapons production) or where there are important government regulations (for example regarding automobile emissions), choice of technology will rest heavily upon a 'dialogue of experts' at all stages of development. Public agencies will certainly be involved as well.[53]

Authorities like Dosi, and Nelson and Winter, argue persuasively that this view of technological change has many advantages over the orthodox neo-classical position.[54] To begin with, it opens up Rosenberg's 'black box' in a sense that it actually provides the beginnings of a realistic explanation of the phenomenon itself, instead of relegating it to the role of an exogenous 'catch-all'. Secondly, it is consistent with what we know about technological changes as these have occurred since the industrial revolution. For example, Rosenberg's interesting cases of older technological vintages coexisting with the newer ones, and indeed improving progressively, is perfectly in keeping with the notion of an older 'paradigm' struggling to maintain its *raison d'être* in the face of an apparently successful newcomer. Thirdly, it permits the introduction of important questions of political economy (like, for example, the role of government agencies) which are evidently germane to science policy issues as these relate to innovation. Finally, it provides some kind of reconciliation between the 'demand pull' and 'science push' theories about the genesis of technological change in modern industrial capitalism. The ideas are by no means complete. For example, there are difficult problems involved in actually defining a technological system. However, for the first time they hold out the hope of being able to explain the behaviour of production units, and indeed national agencies, in so far as these make decisions on technology policy which evidently have very little to do with conventional economic aggregates.

An evolutionary theory of firm behaviour and technological change

In a series of articles in recent years, Nelson and Winter have attempted to weld these new approaches to technological dynamics into a theory of firm behaviour which is essentially biological in character. It is unreal, they argue, to treat firms as if they exist in a world of perfect knowledge and foresight, maximizing objectives functions in an environment of given choice sets and well-defined constraints. By so doing, 'orthodox theory' has little or nothing to say about how or why economic conditions change (for example, about how firms respond to changing market conditions), and therefore about a significant area of economic activity. They go on to claim that their own approach can produce precisely the same theoretical propositions as 'orthodoxy', but with less restrictive and more realistic assumptions about technical conditions and organizational behaviour. For example, firms are categorized as 'profit-seeking' rather than 'profit-maximizing', and 'uncertainty' is explicitly included as an essential feature of economic activity. Finally, they argue, their approach relates more directly to matters of public policy.

Nelson and Winter then proceed as follows: economic change takes place in an 'evolutionary' fashion, where firms are constantly in competition with each other in an unstable environment. Thus, firms are viewed as behaving like organisms constantly under threat and using whatever means are available to perpetuate their existence. Like biological organisms, their 'genes' are institutional 'routines' which define the ways they behave. Some of these 'routines' are 'search routines' which allow institutions to respond (or mutate) in the face of uncertainty and changing external conditions. The successful firms are those which remain in existence and grow. The unsuccessful firms simply go out of business. Economic life is essentially Darwinian.

Nelson's and Winter's ideas have an intuitive appeal. Firms do spend resources on R & D and thereby establish a competitive lead over rivals. They take over other firms (or parts of firms) where this is in their interests. They diversify horizontally and vertically as an insurance against future risks. They clearly do not *just* maximize profits but have complicated sets of goals which themselves change in response to influences both within and outside the firms. Amongst these, security and growth are clearly very important. In short, firms behave in a Schumpeterian manner where time is important and uncertainty very great. One very important determinant of their behaviour is technology and technological change. Nelson and Winter argue that technological choice does not at all take place as a sort of disembodied selection 'from the shelf' but is, on the contrary, programmed into the organic behaviour of the firm. At any point in time, and with respect to any given industry, firms operate on a natural technological trajectory which is partly historically determined and which defines the 'paradigm' for current productive behaviour and R & D for the future. Although this trajectory is common to the industry, each firm has its own room for manœuvre (and competitive edge), determined by its 'in-house' expertise, technical skills, patents, reputation, links to specialist suppliers, and so on. And it is here that the competitive game is played out. Clearly, no firm is going to undertake the enormous costs of moving on to a radically different technological trajectory unless it is entirely convinced that the long-term economic gains are going to be large indeed in relation to the risks. Looked at in this light it is not surprising to find businessmen often responding conservatively to economic signals.

More recently, Nelson and Winter have brought many of these ideas together in a book[55] clearly written to persuade their fellow economists of the conceptual value of an evolutionary approach when modelling innovation and technological change. It is rather a complicated piece of work, unfortunately, and appears to suffer somewhat from the attempt to relate new ideas to prevailing orthodoxy. Hence, while the authors could presumably have developed their ideas further, for example in the direction of general systems theory as McKelvey[56] has done or in that of the behaviour of the firm (see some recent work by Kay,[57] for example), instead they chose to relate their arguments more to the interests of professional economists by means of a complex modelling schema in which all innovation is treated as process innovation.

One important problem, common to many economists, is a clear reluctance to lay down what they expect from a scientific theory and hence how economic modelling and analysis stand up to canons of scientific enquiry. Indeed, the frequent use of the words 'orthodox', 'heterodox' and 'normative' carry distinct overtones of scholastic debate, while it is clear that Nelson and Winter are at pains to distance themselves from more robust critics, like Galbraith.

Kondratiev long waves

Stagnation and recession in the 1970s, combined with associated high levels of unemployment, have revived interest in problems of long-term structural change in advanced capitalist economies, and particularly in long cycles of economic activity. In this interest innovation and technological change play a major explanatory role and again Schumpeter has a clear influence on the discussion. Like Nelson and Winter, the long-wave theorists are dissatisfied with orthodoxy, but their quarrel (and its resolution) takes place at the 'macro' level and with respect to long-term economic growth. For example, Freeman and his colleagues, who in recent years have played a leading role in articulating the notion of long waves,[58] argue the following propositions:

(i) That long waves are statistically discernible.
(ii) That they are closely associated with cycles of innovative activity.
(iii) That during each cycle the relevant technologies are related to each other in a systemic way.
(iv) That, therefore, in so far as long-term structural problems are associated with such cycles (e.g. unemployment and inflation), their resolution requires technology policy as a primary ingredient.

Freeman's starting-point is Schumpeter's treatment of business cycles, articulated in a book of the same name published in 1939. In this book Schumpeter made use of the notion of long waves of economic activity, first identified by van Gelderen[59] in 1913 and subsequently given an extended treatment by the Marxist economist Kondratiev[60] in 1925. These superimposed themselves on the shorter 'trade' or 'juglar' cycle, normally associated with cyclical activity and lasting around five to seven years. In contrast, the typical Kondratiev wave was observed to last fifty years or so, during which time the economic system ran the full gamut of boom, recession, depression and back to boom again.

Schumpeter's position was not only that such waves were historically discernible, they were also clearly and primarily associated with bursts of innovative activity and the rise and decline of particular industrial sectors. The first 'Kondratiev' (or 'K') was linked to the emergence of the sequence of industries leading the Industrial Revolution—cotton textiles, coal and iron. The second K was generated by the railway boom of the period *c.*1850–70, while the third involved the development of automobiles, electricity and radio in the twenty or so years before World War I. The years following each K were those of long-term depression typified by the major depressions of 1870–90 and the inter-war period. To these three Ks, Freeman adds a fourth, associated with the years of fast economic growth in Europe, the United States and Japan between 1950 and 1970, and led by industries like bulk chemicals, pharmaceuticals, electronics, aerospace and nuclear power.

The way each cycle unfolds is as follows: to begin with, a major technology is innovated and takes root. Often, the groundwork for it has already taken place in a previous K, in terms of inventions and development of ancillary technologies. However, the major technology itself is a many-headed hydra involving whole clusters (swarms) of new innovations 'affecting processes, components, sub-systems, materials and management systems, skills and finance, as well as the products themselves'.[61] Moreover, there are many different product types involved. The railway boom was not simply concerned with the building of steam engines and the laying of track. It also involved the development of signalling systems, the building of rolling stock, the manufacture of appropriate forging and machine-tool facilities, the establishment of stations and marshalling yards, and the inculcation of the whole vast range of technical and managerial skills required to run a railway system.

Once a boom gets under way the explosive growth creates a climate of investment optimism which is self-fulfilling. New and related products are introduced, demand increases, as do employment and the general level of wages. Eventually, however, the 'long boom' comes to an end as the new technological system becomes fully integrated within the economy. As productive potential falls off, new investment drops to zero and what is sometimes called a 'multiplier/accelerator' mechanism comes into operation, bringing with it rapidly diminishing incomes and demand. During this phase, price competition between productive units becomes particularly fierce as innovative activity becomes focused upon cost-reducing forms (process innovation) rather than those of creating new products (product innovation). Nevertheless, many firms go bankrupt and the overall rate of unemployment increases. Eventually, once a 'low point' is reached, a radically new technology appears upon the scene whereupon a new Kondratiev cycle begins.

Freeman and his colleagues have taken these ideas of Schumpeter and subjected them to empirical and theoretical scrutiny using new evidence on patents and innovation for the period 1900 to the present day. They conclude that although long waves do provide a useful 'heuristic' within which to analyse industrial capitalism, they should by no means be viewed as a form of crude 'technological determinism' whereby a new major innovation will suddenly appear, like a phoenix out of the ashes of a deep depression. Although there are regularities, technological changes are extremely uneven over time and geographically (i.e. between regions and countries). There are also apparent differences between successive Ks. Thus, the recent long wave is characterized by much greater institutional rigidities than previous ones, with consequently high rates of inflation coexisting with recessionary unemployment. Moreover, there is a very real possibility that service sector employment may not be the ever-ready 'sponge' to mop up redundancy in manufacturing employment that it has been in the past.[62] These features, plus the surprisingly rapid internationalization of technology to the NICs, make it at least arguable that any long-term resurgence into a new (fifth) Kondratiev (fuelled, say, by the microprocessor or biotechnology) will have very different social implications from anything that has transpired in previous periods, particularly in the areas of occupational employment and income distribution. In consequence, Freeman's principal conclusion is that deliberate technology policies associated with greater public expenditures may be necessary if countries like Britain are to successfully 'climb on' to the next long wave. Certainly, traditional 'demand management' policies by themselves are unlikely to act as an effective means for promoting economic development in the increasingly competitive (and international) environment of the 1980s and 1990s.

Galbraith and the modern industrial state

Up to now in this section, most of the 'radical' conceptual developments I have surveyed have tended to focus directly on the essential fabric of technological changes as these take place within modern industrial capitalism. While the arguments deployed are often critical of established economic orthodoxy, they are not intended as the building blocks in any larger social critique. In this respect John Kenneth Galbraith occupies rather a unique position in so far as he places very strong emphasis upon the influence of science and technology on modern industry and government and hence upon practically every aspect of our daily lives—an influence he clearly finds profoundly depressing. He represents therefore a throwback, as it were, to the older classical tradition of political economy where it was expected that social scientists would involve themselves more deeply in grand matters of public debate, using their analytical tools as a means towards a larger end.

The key to much of Galbraith's thinking is contained in the *New Industrial State*[63] and relates to the differentiation of economic production which is associated with the history of technological changes, a point also brought out by Cooper[64] and Rosenberg.[65] Using the automobile industry as his 'case study', he shows how the extent to which scientific knowledge may be applied systematically to economic production is a function of the degree of specialization of the production system. But specialization also brings along with it need for co-ordination and control, greater complexity of production, heightened risks and the need for much larger sums of investment capital. Nowadays, the development lead time for a new model is often five years or more. In Henry Ford's time it was a matter of months. What we have then are series of *inevitable* consequences of the use of modern technology which are programmed into the very fabric of much of economic production. The obverse of the coin of the high productivity effects of science and technology is the sheer scale and inflexibility of the modern industrial combine. Nor is this all. The large corporate organization is itself a power centre which is self-justifying. Profit maximization is no longer the main motivation since the bureaucratic nature of the modern corporation has shifted power from the owners (the shareholders or capitalists) to the managerial élite (the technostructure) who run things. Shareholders are kept in a state of quiescence with a minimum level of profits sufficient to permit moderate dividends, thereby allowing the technostructure to pursue corporate goals of a more complex variety but of which 'growth' and 'survival' are clearly important elements. So far as possible, the external environment has to be kept firmly under control. Sources of supply are ensured through long-term contracts and, where necessary, backward integration. Consumers are 'programmed' into the acceptance of products by means of advertising, sales promotion, market research and similar devices. In this way the traditional role of consumer sovereignty, so beloved of neo-classical orthodoxy, is tempered to the exigencies of modern productive forces. What Galbraith has called 'producer sovereignty' takes over.

In the later book, *Economics and the Public Purpose*,[66] Galbraith tries to put his argument within the broader context of the political economy of advanced capitalism which he sees as moving inexorably towards an Orwellian system of bureaucratic privilege and power. Modern economic production, he argues, is conducted within two broad types of sector which stand in a relationship of dynamic inequality to each other—the 'Planning System' and the 'Market System'. The Planning System is comprised of large corporations, Departments of State and powerful trade unions, all of which are heavily bureaucratized. It is characterized by advanced technology, monopolistic control over markets and hence the ability to control prices and wages. As with the corporations analysed in the *New Industrial State*, organization and producer sovereignty reign supreme, and there are close organic links between its various components.

Conversely, the Market System is composed of small entrepreneurial businesses dealing with personal services, agriculture and geographically dispersed functions. The labour force is often not unionized, or only weakly, so there is little access to the State bureaucracy and competitive market conditions tend to obtain along with, correspondingly, a substantial degree of consumer sovereignty. The relations between the two systems are such that the Planning System constantly gains at the expense of the Market System in terms of resources, incomes and power. Since competitive pressures are relatively absent, the Planning System can hike wages and prices during inflationary periods and can protect itself conversely during periods of recession. The Market System, on the other hand, cannot protect itself in this way and hence bears the full brunt of cyclical fluctuations. It follows, of course, that conventional counter-cyclical policy will not work in the ways intended, at least not so far as the Planning System is concerned. A monetary 'squeeze' can be absorbed by large corporations which have their own sources of finance, while tax and expenditure fiscal changes can either be absorbed directly or passed on to the Market System which tends therefore to suffer disproportionately. In this way power, wealth and privilege are allocated in a manner which has little to do with 'marginal productivity', but rather is a function of increasingly rigid institutional patterns.

But Galbraith goes further than merely suggesting a new model of social development, arguing that neo-classical economics, as a professionalized activity, actively colludes in the prevailing set of power relations by postulating a body of ideas (or doctrine) which describes a world totally at variance with reality—a world in which the *whole* economic system is subject to the discipline of the market. The fact that neo-classical 'market forces' are simply not capable of functioning over large areas of economic activity is treated with a heavy, if rather embarrassed, silence. Finally, his model postulates a form of class conflict which cuts across traditional Marxist categories. No longer is the antagonism one between homogeneous 'workers' and 'capitalists'. It is between 'organized' workers and 'unorganized' workers, between 'organized' industry and 'unorganized' industry. Even within the various public services (like medical care, for example), Galbraith would not see in the British National Health Service an institution devoted to the care of the masses. Rather, he would regard it as a centre of privileged power, dominated by a (male) establishment with a 'high technology' approach to medical science, closely linked with large and powerful drug companies, and where the interests of the health 'consumer' play second fiddle to the certificated 'expertise' of the 'producer'. More generally, the close tie-in between scientific expertise and bureaucratic privilege in health, welfare, education, energy supply, and many other areas of modern economic and social activity, has produced a new form of bureaucratic centralism becoming ever more remote from popular understanding and control. Even at Central Government level, the various 'Departments of State' have great difficulty in keeping abreast of each others' activities, particularly where 'high technology' is concerned. In countries like Great Britain, where there is a long tradition of departmental autonomy and responsibility, this is much more of a problem than in countries like Japan, say, where departments of state tend to be more organic and functionalist.

Not everyone finds it easy to accept Galbraith's pessimism and there is more than a hint of technological determinism in his writings. Moreover, on the whole the depth of his 'diagnosis' is not matched by the acuity of his policy prescription where he appears to place a lot of faith on exhortation and on the good sense of the 'experts', whom he calls the Educational and Scientific Estate. Finally, although he has argued for the conscious use of state power to mitigate the pervasive tendencies to socio-economic inequality, it is not easy to see how powerful bureaucratic entities will connive at their own emasculation. Nevertheless, his views need more confrontation than they appear to have received. One has to suspect that economic orthodoxy is more than a little afraid to take him on.

Some concluding comments

This essay has comprised a broad survey of modern developments in the study of technological innovation. Newcomers to the subject may well feel by this stage that the sheer range and complexity of this discussion is hardly matched by the resultant light shed on social policy. This, I am afraid, is a well-known failing of social science, but what do we expect a theory in the social sciences to tell us? One response is that of robust empiricism, like that of Pavitt and his colleagues who appear to argue that until better data are collected, it is premature to theorize. The problem with this view is that the collection of 'more' and 'better' evidence requires of necessity a set of prior determinant theories. Indeed, Pavitt himself tacitly admits this in a recent article in which he defends his re-classification of industry into three 'genotypes' which are characterized by distinctly different innovative behaviour[67]—a classification which he clearly feels will lead to better evidence, and hence better theory in a continuous virtuous spiral. To take the biological analogy further, classification of a species (of sub-species) must always depend upon a higher level set of theories which provides the basis for such a classification.

At the risk, therefore, of premature speculation, I should like to close this essay with a few remarks which are apposite in this respect. On the diagnostic side it is clear that neo-classical economic analysis can only be 'fitted to' the experience of technological change in modern economic systems with extreme difficulty and the result is often more than a little artificial. I have argued elsewhere[68] that this is really only to be expected since economics as a discipline was not designed to handle either complex socio-economic change *through time* or the *interdisciplinary problems* which result. We have seen in this essay that Freeman, and a few others, are trying to break out of this 'closed system' by casting around for suitable metaphors which stand some chance of acting as a more promising set of conceptual organizers for the analysis of technological change. But they are up against two major obstacles. One is the tenacity with which the economics profession as it is currently institutionalized attempts to maintain its intellectual hegemony. The other is a certain reluctance to lay down the canons of scientific enquiry against which a fruitful conceptual debate may take place since, in the absence of this, discussion tends to degenerate into dogmatic controversy. My own strong suspicion is that the weakness of economic analysis in this field is largely a function of its essentially deductive and Cartesian trappings. It is no accident that many of the more modern approaches are organic and evolutionary in character, comprising a way of viewing the phenomenon of technological change which is arguably not only more realistic but has also a much greater chance of relating directly to the needs of policy makers.

Notes

1. There are a number of review articles in this field, written from different points of view. Particularly useful are: C. Kennedy and A. Thirlwall, 'Technical Progress: A Survey', *Economic Journal*, **82**, (1972), pp. 11–72; N. Rosenberg, 'The Historiography of Technical Progress', in N. Rosenberg, *Inside the Black Box: Technology and Economics*, Cambridge University Press, 1982, pp. 3–33; and M. Fransman, 'Conceptualizing Technological Change in the Third World: An Interpretive Survey', *Journal of Development Studies* (forthcoming). Also very useful as source texts are A. K. Sen (ed.), *Growth Economics*, Harmondsworth, Penguin, 1970; N. Rosenberg (ed.), *The Economics of Technological Change*, Harmondsworth, Penguin, 1971; and C. Freeman, 'Economics of Research and Development', in I. Spiegel-Rösing and D. de Solla Price (ed.), *Science, Technology and Society: A Cross-Disciplinary Perspective*, London, Sage, 1977, Chap. 7.
2. Many of the writings of these authors will be cited below.
3. A. K. Sen (ed.), *Growth Economics*, Harmondsworth, Penguin, 1970, p. 9.
4. See House of Lords Select Committee on Science and Technology, *Engineering Research and Development*, Session 1982–83, London, HMSO, 1984, Vol. 1, for a useful set of statistics on these points.
5. R. Harrod,'An Essay in Dynamic Theory', *Economic Journal* **49** (1939), pp. 14–33. Also in Sen, op. cit., note 3, 43–64.
6. R. M. Solow, 'Technical Change and the Aggregate Production Function', *Review of Economics and Statistics*, **39** (1957), pp. 312–20. Also in A. K. Sen, op. cit., n. 3, pp. 401–19.

7. M. Abramovitz, 'Resources and Output in the U.S. Since 1870', *American Economic Review, Pap. Proc.*, **46** (1956), pp. 5–23.
8. I have surveyed some of this early literature in my doctoral thesis, *The Techno-Economic Relationship Between Industry and the Scientific Infrastructure*, unpublished Ph.D. dissertation, University of Edinburgh, 1981.
9. This point is made by C. Freeman in 'Research and Development in Electronic Capital Goods', *National Institute Economic Review*, No. 34 (1965), pp. 29–57.
10. Although it is only fair to point out that Freeman has always argued that R & D does indeed contribute to overall economic growth, through this process, and others.
11. E. Mansfield, *The Economics of Technological Change*, London, Longman, 1968.
12. J. R. Minasian, 'The Economics of Research and Development' in U.S. National Bureau of Economic Research, *The Rate and Direction of Inventive Activity*, Princeton, Princeton University Press, 1962, pp. 93–142.
13. See, for example, I. Illich, *Tools for Conviviality*, London, Fontana, 1975 and F. Capra, *The Turning Point*, New York, Simon and Schuster, 1982.
14. J. Irvine and B. Martin, *Foresight in Science: Picking the Winners*, London, Frances Pinter, 1984.
15. E. Denison, 'United Sates Economic Growth' *Journal of Business*, **35** (1962), pp. 109–21. Also in N. Rosenberg (ed.), op. cit., note 1, pp. 363–81.
16. See, for example, Z. Griliches, 'The Sources of Measured Productivity Growth: U.S. Agriculture, 1940–60', *Journal of Political Economy*, **71** (1963), pp. 331–46.
17. G. J. Stigler, 'Economic Problems in Measuring Changes in Productivity' in National Bureau of Economic Resarch, *Output, Input and Productivity Measurement*, Princeton, Princeton University Press, 1961, 47–55.
18. See, for example, Denison, op. cit., note 15.
19. Z. Griliches, 'Hybrid Corn: An Exploration in the Economics of Technological Change', *Econometrica*, **25** (1957), pp. 501–22.
20. Rosenberg, op. cit., note 1, p. 25.
21. See C. Freeman, 'Economics of Research and Development', in I. Spiegel-Rösing and D. de Solla Price (eds), op. cit., note. 1, p. 24, for a discussion of these points of criticism.
22. N. Rosenberg, 'Science, Invention and Economic Growth', in N. Rosenberg, *Perspectives on Technology*, Cambridge, Cambridge University Press, 1976, p. 260.
23. J. Schmookler, *Invention and Economic Growth*, Cambridge, Harvard University Press, 1966, p. 208. Quoted in R. Rothwell and W. Zegveld, *Reindustrialization and Technology*, Harlow, Essex, Longmans, 1985, p. 24.
24. B. Hessen, 'The Social and Economic Roots of Newton's *Principia*', in N. Bukharin (ed.), *Science at the Crossroads*, London, Kniga, 1931; reprinted London, Frank Cass, 1971, pp. 151–212.
25. See, for example, V. Walsh *et al.*, 'Invention and Innovation in the Chemicals Industry: Demand Pull or Disovery Push', *Research Policy*, **13** (1984), pp. 211–34, and D. C. Mowery and N. Rosenberg, 'The Influence of Market Demand upon Innovation: a Critical Review of Some Recent Empirical Studies' in N. Rosenberg, *Inside the Black Box*, op. cit., note 1, p. 235.
26. Rosenberg, op. cit., note 22, p. 278.
27. M. Posner, 'International Trade and Technical Change', *Oxford Economic Papers*, **13** (1961), pp. 323–41.

28. S. Hirsch, *Location of Industry and International Competitiveness*, unpublished Ph.D. dissertation, Harvard Business School, 1965.
29. G. C. Hufbauer, *Synthetic Materials and the Theory of International Trade*, London, Duckworth, 1966.
30. R. Vernon, 'International Investment and International Trade in the Product Cycle', *Quarterly Journal of Economics*, **80** (1966), pp. 190–207.
31. This is the economist's definition, which should be contrasted with the engineer's, which tends to view a 'technology' more in terms of a collection of machines and assemblies.
32. W. Leontief, 'Domestic Production and Foreign Trade; the American Capital Position Re-examined' in W. Leontief, *Input-Output Economics*, New York, Oxford University Press, 1966, pp. 68–99.
33. The original source is J. Schumpeter, *The Theory of Economic Development*, New York, Oxford University Press, 1961, see Chap. 4. However, a good summary discussion is contained in C. M. Cooper, 'Science, Technology and Development', *Economic and Social Review*, **2** (1971), pp. 165–89.
34. See C. M. Cooper (ed.), *Science, Technology and Development*, London, Frank Cass, 1973, Chap. 1.
35. See Fransman, op. cit., note 1.
36. Rosenberg, op. cit., note 1, p. 107.
37. J. Schumpeter, *Capitalism, Socialism and Democracy*. London, Allen and Unwin, 1976.
38. J. K. Galbraith, *The New Industrial State*, Harmondsworth, Penguin, 1971, Chap. 1.
39. F. Scherer, *Industrial Market Structure and Economic Performance*, Chicago, Rand McNally, 1970.
40. C. Freeman in Spiegel-Rösing and de Solla Price (eds), op. cit., note 1, p. 257.
41. Rosenberg, op. cit. note 1, pp. 141–59.
42. See, for example, E. Mansfield, 'Technical Change and the Rate of Imitation', *Econometrica*, **29** (1961), pp. 741–66, and S. Davies, *The Diffusion of Process Innovation*, Cambridge, Cambridge University Press, 1979. For a critical review, see L. Soete, 'International Diffusion of Technology, Industrial Development and Technological Leapfrogging', *World Development* (forthcoming).
43. The material from these is summarized in C. Freeman, *The Economics of Industrial Innovation*, Harmondsworth, Penguin, 1974.
44. D. de Solla Price, 'Is Technology Historically Independent of Science?', *Technology and Culture*, **6** (1965), p. 553.
45. J. Jewkes, D. Sawers and R. Stillerman, *The Sources of Invention*, London, Macmillan, Rev. edn., 1969.
46. E. Mansfield *et al.*, *Research and Innovation in the Modern Corporation*, New York, Norton; and London, Macmillan, 1971.
47. See H. Stead, 'The Costs of Technological Innovation', *Research Policy*, **5** (1976), pp. 2–9.
48. Freeman, in Spiegel-Rösing and de Solla Price (eds), op. cit., note 1, p. 251.
49. See, for example, M. Gibbons and R. Johnston, 'The Role of Science in Technological Innovation', *Research Policy*, **3** (1974), pp. 220–42.
50. Rosenberg, *Perspectives in Technology*, op. cit., note 22, pp. 191–210.
51. See, for example, G. Dosi, 'Technological Paradigms and Technological Trajectories, a Suggested Interpretation of the Determinants and Directions of Technical Change', *Research Policy*, **11** (1982), pp. 147–62.

52. See Freeman, op. cit., note. 43, Chap. 7, for a more detailed discussion of this point.
53. For a useful discussion of US experience, see D. Nelkin, 'Technology and Public Policy', in Spiegel-Rösing and de Solla Price (eds), op. cit., note 1, pp. 393–441.
54. See Dosi, op. cit., note 51, and R. Nelson and S. Winter, 'In Search of a Useful Theory of Innovation', *Research Policy*, **6** (1977), pp. 36–77.
55. See R. Nelson and S. Winter, *An Evolutionary Theory of Economic Change*, Cambridge, Harvard University Press, 1982.
56. B. McKelvey, *Organisational Systematics*, London, University of California Press, 1982.
57. N. Kay, *The Evolving Firm*, London, Macmillan, 1982.
58. See C. Freeman, J. Clark and L. Soete, *Unemployment and Technical Innovation*, London, Frances Pinter, 1982.
59. J. van Gelderen, 'Springvloed: Beschouwingen over industriële ontwikkeling en prijsbeweging', *De Niewe Tijd*, **18**, (1913), 4, 5 and 6. Quoted in Freeman *et al.*, op. cit., note 58, p. 19.
60. N. Kondratiev, 'The Major Economic Cycles', reprinted in *Lloyds Bank Review* 129 (1978), pp. 41–60.
61. C. Freeman, *The Guardian*, 30 August 1983, p. 16.
62. For an account of this argument see J. Gershuny, *After Industrial Society*, London, Macmillan, 1978.
63. Galbraith, op. cit., note 38.
64. Cooper, op. cit., note 34.
65. Rosenberg, op. cit., note 22, Chap. 7.
66. J. K. Galbraith, *Economics and the Public Purpose*, London, André Deutsch, 1972.
67. K. Pavitt, *Patterns of Technical Change—Evidence, Theory and Policy Implications*, SPRU Papers in Science, Technology and Public Policy, 1983, p. 6.
68. N. G. Clark, *The Political Economy of Science and Technology*, Oxford, Blackwell, 1985, Chap. 9.

PART I

Economics and Industrial Innovation

1 'Chips' and 'Trajectories': how does the semiconductor influence the sources and directions of technical change?

Keith Pavitt

Introduction

In this essay, I propose a framework to explain what the 'chip', or the micro-electronics revolution based on the semiconductor, is doing to the sources and directions of technical change in the industrially advanced countries. Christopher Freeman's own research style and achievements suggest why and how such an essay should be written.

First, he has always stressed the importance of describing and understanding present and likely future patterns of technical change.[1] In the 1970s, he jointly led a programme exploring the links between economic growth and global resource depletion, concluding that continuous technical change and social adaptation were essential if the former were to be achieved without the latter.[2] In the 1980s, he and his colleagues have built on the work of Schumpeter, in order to explore the implications of the fundamental and pervasive innovation that is the semiconductor, for trends in the level of economic activity in general, and for employment levels and skills in particular.[3] Like him, I shall assume that the semiconductor does have a significant impact on the rate and direction of technical change in a wide variety of sectors. However, I shall concern myself with the impact on the type of firm that will make innovations, and on the nature of the innovations themselves.

Second, Freeman has always argued that the sources, the rate and the direction of invention and innovation cannot be explained by, or derived from, some simple, general rule. After being asked in the late 1960s to contribute to the debate about the role of small firms in making innovations, he began a painstaking programme of data collection on the sources of more than one thousand significant innovations introduced into the United Kingdom since 1945. He was thereby able to show the considerable variance among sectors in the relative contributions of small and large firms.[4] During the 1970s, J. Townsend and others built up this data, so that in 1981 it comprised more than two thousand, covering more than half of British manufacturing.[5] For each innovation, information was collected on the main knowledge sources, the sectors of production and main use of each innovation, and on the size and principal sector of activity of the innovating firm. These data have enabled this writer to describe and explain intersectoral differences in the sources and directions of technical innovation,[6] as well as to define some of the more general characteristics of technology, innovation and technical change.

These sectoral patterns (or 'trajectories') of technical change are the starting-point of this essay. After briefly describing and explaining them, I suggest tentative hypotheses about the way in which the trajectories are influenced by developments in semiconductor technology. Where possible, I compare these hypotheses with empirical evidence, before reaching some conclusions for theory and for policy.

Technological trajectories

Empirical research shows that technological knowledge in the modern economy has two interconnected properties: it is mainly specific in application, and cumulative in development. Its specificity to particular applications is reflected in three characteristics: first, the heavy concentration of firms' innovation—generating expenditures on development and on production engineering activities, both of which are specific to one product and one production process;[7] second, the heavy reliance of the typical innovating firm on knowledge that is not public, but specific to itself or to other firms in technologically related lines of business; third, the considerable expenditures on technological assimilation and adaptation that are typically incurred when transferring technology from one firm to another, or even from one place to another.[8] Given this specificity in technological knowledge and skills, firms do not make a generalized search when deciding where to move next technologically, but explore zones that use technological knowledge similar to that they already know. Firms at different technological starting-points therefore follow different and cumulative technological trajectories. These can be observed in sectoral patterns of firms' innovative activities, as reflected in either significant innovations made, or R & D resources spent.[9]

Nelson and Winter[10] have proposed three factors that determine technological starting points and trajectories: the available sources of technology, the requirements of users, and the possibilities open to the innovating firm to benefit more than its competitors from any innovations on which it spends resources. This formulation, together with the data compiled by Townsend *et al.*[11] on significant British innovations, has enabled this writer to describe and explain sectoral trajectories. Firms can be divided into three categories: supplier-dominated, production-intensive and science-based. Table 1.1 describes their typical core sectors, as well as the nature, determinants and measured characteristics of their technological trajectories.

Table 1.1 Sectoral technological trajectories: determinants, directions and measured characteristics

Category of firm		Typical core sectors	Determinants of technological trajectories			Technological trajectories	Measured Characteristics			
			Sources of technology	Type of user	Means of appropriation		Source of process technology	Relative balance between product and process innovation	Relative size of innovating firms	Intensity and direction of technological diversification
(1)		(2)	(3)	(4)	(5)	(6)	(7)	(8)	(9)	(10)
Supplier-dominated		Agriculture Housing Private services Traditional manufacture	*Suppliers* Research & extension services Big users	Price-sensitive	Non-technical (e.g. trade marks, marketing, advertising, aesthetic design)	Cost-cutting	Suppliers	Process	Small	Low vertical
Production-intensive	Scale-intensive	Bulk materials (steel, glaze) Assembly (consumer durables & autos)	*PE* Suppliers R & D	Price-sensitive	Process secrecy & know-how Technical lags Patents Dynamic learning economies	Cost-cutting (Product design)	In-house suppliers	Process	Large	High vertical
	Specialized suppliers	Machinery Instruments	*Design & Development* Users	Performance-sensitive	Design know-how Knowledge of users Patents	Product design	In-house suppliers	Product	Small	Low concentric
Science-based		Electronics electrical Chemicals	*R & D* Public science PE	Mixed	R & D know-how, patents Process secrecy & know-how Dynamic learning economies	Mixed	In-house suppliers	Mixed	Large	Low vertical High concentric

Note: PE – Production Engineering Department.
Source: K. Pavitt, 'Sectoral Patterns of Technical Change: Towards a Taxonomy and a Theory', *Research Policy*, **13** (1984), pp. 343–73.

Supplier-dominated firms make very little contribution themselves to either their product or their process technology. They can be found mainly in traditional sectors of manufacturing like textiles, in agriculture, in house building, and in many professional, financial and commercial services. Most innovations come from suppliers of equipment and materials, although in some cases contributions are made by government-financed research and extension services, by large customers, or by the relatively few large firms in the sector. Technological trajectories are defined in terms of cost-cutting, based on what is offered by suppliers.

Over time, some firms will evolve from the supplier-dominated to the *production-intensive* category. In 1776 Adam Smith described one of the mechanisms of this evolution: an increasing division of labour and simplification of tasks in production, resulting from an increased size of market, and leading to an increasing substitution of machines for labour. The pressures and incentives to exploit scale economies are particularly strong in what I call scale-intensive firms selling to largely price-sensitive users: those producing standard bulk materials through continuous processes and those producing durable consumer goods and vehicles. In both cases, production processes have become increasingly large, complex and interdependent. Their smooth operation cannot be taken for granted and the costs of failure in any one part of the production system are considerable. Satisfactory operation has therefore come to depend on increasingly professionalized 'production engineering' or 'process engineering' departments, which themselves become an important source of technical change in production processes, and related machinery and instruments.

These scale-intensive firms live in symbiosis with specialized firms supplying production machinery and instrumentation, and who have different technological trajectories from their customers. Given the scale and the interdependence of the production systems to which they contribute, the costs of poor performance of their products are considerable. The technological trajectories of specialized suppliers are therefore strongly focused on the performance and reliability of their products. Specialized suppliers benefit from close and continuous contact with their customers, who often pass onto them skills, information on operating experience, improvements made to equipment in use, and even resources for the design and testing of new equipment. At the same time, each customer benefits from suppliers who have assimilated information and improvements from a large number of users.

In the nineteenth century, *science-based* firms began to emerge, as a result of key scientific discoveries: electromagnetism, radio waves and transistor effects contributed to what is now the electrical and electronics industry, while chemical synthesis and biological synthesis have been the basis of what is now the chemical industry. These key discoveries, and related R & D activities in universities and firms, have been determining influences on the firms' technological trajectories. The pervasive and varied range of applications growing out of these science-based techniques have both enabled rapid growth for successfully innovating firms, and dictated varying emphases on product or process innovation. In bulk synthetic materials and standard consumer products, the same pressures for cost-cutting and increased scale of production have existed as in the production-intensive sectors, whilst in pharmaceuticals, other fine chemicals and in electrical and electronic machinery and equipment, there has been greater pressure for product reliability and performance.

The semiconductor and sectoral directions of technical change

From this analysis, it is clear that the *directions* of technological accumulation in firms are conditioned by the opportunities opened up by fundamental technologies. As Perez has argued,[12] such technologies are defined by the potential that they offer for cost reduction and new products in a wide range of applications. Thus, science-based firms have benefited principally from techniques emerging from research in physics and chemistry; whilst firms in the production-intensive categories have benefited from technologies that have increased the range of potentially useful mechanical products that can be made, and the possibilities of exploiting scale economies in producing them: in particular, from steel-making, electricity, petroleum and the internal combustion engine.

Similarly, future directions of technical change in firms will be influenced by both their existing technological trajectories and the following specific characteristics of semiconductor technology: first, that it grows out of electronics technology; second, that it contributes to two primary technical functions—information processing, and monitoring and control; third, that both these technical functions have improved at a prodigious rate in the past thirty years, as a result of continuous and rapid improvements in product design, allied to steep dynamic learning economies in the production of semiconductor-based circuitry.[13]

Given these characteristics, we would expect semiconductor technology—like earlier science-based technologies—to have pervasive effects on process technologies in all sectors. Monitoring and control technologies will be adopted particularly rapidly in those sectors where production technologies are large-scale and interdependent, and where firms have the in-house competence to use, to specify and, if necessary, to develop the appropriate production/process control systems: in other words, in firms in our science-based and production-intensive categories. As a consequence, we would also expect improvements in monitoring and control systems to offer opportunities for product innovation more generally in sectors that make production equipment. At the same time, semiconductor technology will help continue the past trajectory of the electronics/electrical sector in creating ample opportunities for product innovations to be adopted in manufacturing, services and households.

Data collected by Northcott, Rogers and Zeilinger on the use of microelectronics in more than one thousand establishments in British manufacturing in 1981 confirm these expectations.[14] Table 1.2 shows, for each two-digit sector, the proportion of establishments in the sample that had used microelectronics in processes and in products. The third column shows that the sectors that we have defined as *science-based* and *production-intensive* (food and drink, chemicals and metals, mechanical engineering, electrical and instrument engineering, vehicles) have above average use of microelectronics in processes, whilst those that we define as *supplier-dominated* (textiles, clothing and leather) have below average use. Furthermore, the second column shows that the opportunities for product innovations are heavily cononcentrated in electronics and instrument engineering, but are also above average in mechanical engineering and vehicles, again in line with our expectations.

The semiconductor and the sectoral sources of technical change

Thus, with the partial exception of chemical firms, the spread of semiconductors will not change fundamentally the intersectoral differences in the balance between product and process innovations made by firms, or in the rate of adoption of advanced process technology. But what effect will it have on the sectoral *sources* of semiconductor-related technology? Whilst we can expect an increasing use of electronic skills and electronics products in a wide range of sectors, our analysis of the cumulative nature of technical change suggests that the sectoral sources of electronics-based innovations in production-intensive and supplier-dominated firms will change only marginally, if at all.

Production-intensive firms

We would expect that the main effect of semiconductors in production-intensive firms will be the augmentation of the already existing process-dominated technological trajectories, through the application of monitoring and control technology. As with earlier mechanical and electromechanical production technologies, we would expect significant contributions to their development by both large firms in the user sectors, and small specialized firms supplying equipment incorporating monitoring and control functions.

Evidence from the SPRU data bank on the characteristics of British

Table 1.2 Use of microelectronics by industry

Industry	No. of establishments	Product users	Process users	All users	Non-users	All
		(percentages of the establishments in each industry)				
Food & drink	(125)	0	56	56	44	100
Chemicals & metals	(134)	0	51	51	49	100
Mechanical engineering	(165)	29	43	55	45	100
Electrical & instrument engineering	(133)	58	60	76	24	100
Vehicles	(92)	16	51	54	46	100
Other metal goods	(99)	4	40	40	60	100
Textiles	(97)	0	31	31	69	100
Clothing & leather	(92)	1	21	21	79	100
Paper & printing	(99)	0	52	52	48	100
Other manufacturing	(164)	5	40	40	60	100
TOTAL	(1,200)	13	45	49	51	100

Source: J. Northcott, P. Rogers and A. Zeilinger, *Microelectronics in Industry: Survey Statistics*, London, Policy Studies Institute, 1982.

innovations and innovating firms in instrument engineering between 1945 and 1980 confirms that this is the case. Instrument engineering is the product group that best represents products embodying electronics-based functions for monitoring and control. Significant innovations have mainly been made by both specialized supplier firms and by larger firms, some of which are themselves users of the innovations. As Table 1.3 shows, the sources and directions of innovations in instrument engineering are similar to those in mechanical engineering. Innovating firms with their principal activities in instruments are relatively small (columns 4–6), technologically specialized (column 2) and with a high proportion of their innovations used in other sectors, i.e. product innovations (column 3); at the same time, a relatively high proportion of innovations are made by relatively large firms (including users) principally engaged in other sectors (columns 8–10). Furthermore, Table 1.4 shows in column 4 that users of instrument engineering in process technology are fairly evenly spread across manufacturing, accounting overall for 13 per cent of the process innovations adopted. Column 2 shows that 6 per cent of the innovations produced by manufacturing firms, other than those principally in instruments, use instrument innovations, the percentages being higher in firms typified by continuous processes (iron and steel, cement and glass, paper), as well as in shipbuilding and in electronics and electrically-based firms.

Similar sources of innovation can also be observed in the development of major electronics-based technologies in manufacturing assembly: computer aided design (CAD), numerically controlled machinery (NCM), and robots, where a large number of studies describing and explaining their emergence and diffusion have been completed in the last two years.[15] They all show the major role of users of the technologies in their development. In the 1950s and early 1960s, the major stimuli came from the US Federal Government, for the machining of parts for the high performance products of the aerospace industry; there were similar but smaller programmes in France and the United Kingdom. By the late 1960s, the locus of technological activity began to shift towards the electrical/electronics, automobile and shipbuilding sectors, and towards Western Europe and Japan. According to Gönenc,[16] automobile firms have played a major role in the development and the diffusion of the technologies, given their size, technological resources, and central position in a range of metalworking and metal-using technologies. According to Carlsson, users have been the main driving force in technological change in NCM.[17] The same pattern emerges from a recent Japanese study of the introduction of assembly-based automation.[18]

Users' major involvement results from systemic interdependence in production, and from differentiated technological requirements.

Table 1.3 Characteristics of innovation firms and innovations in instrument engineering and mechanical engineering

Sector	% of innovations produced by firms principally in the sector that are:		Size distribution of employment of innovating firms principally in the sector (Rows add up to 100%)			Innovations in the sector that are made by firms with their principal activities in other sectors			
	Innovations in other product	Used in other sectors	10,000+	1,000–9,999	1–999	As % of all innovations in sector	Size distribution of firms		
							10,000+	1,000–9,999	1–999
(1)	(2)	(3)	(4)	(5)	(6)	(7)	(8)	(9)	(10)
Instrument engineering	19.9	81.4	24.6	21.4	54.0	54.8	54.1	13.8	32.0
Mechanical engineering	16.0	82.5	24.3	36.9	38.8	31.9	67.3	15.2	17.5
All sectors in the sample	31.5	64.0	53.1	21.9	24.9	31.4	53.1	21.9	24.9

Source: SPRU Databank on British innovations since 1945.

Table 1.4 Instrument innovations produced and used in manufacturing sectors other than instrument engineering

Sector	Instrument engineering innovations as % of all innovations produced by firms principally in the sector		Instrument engineering innovations as % of all innovations used in the sector	
	% of all innovations produced	All innovations produced	% of all innovations used	All innovations used
(1)	(2)	(3)	(4)	(5)
Food and drink	1.3	78	10.3	68
Chemicals	4.8	78	10.3	71
Metal production	11.9	143	19.2	130
Mechanical engineering	6.5	536	20.1	169
Electronics and electrical	15.2	343	12.0	167
Shipbuilding	12.4	89	10.0	90
Vehicles	3.2	158	8.1	221
Textiles	2.6	77	12.7	377
Leather and footwear	2.0	50	4.4	45
Glass and cement	11.5	87	19.0	63
Paper	11.6	43	10.3	39
Total manufacturing (excluding instruments)	6.0	1,946	13.4	1,452

Source: SPRU Databank on British innovations since 1945.

Symbiotic relationships exist between large-scale users and specialized suppliers of production equipment, with operating experience, specifications, designs and skills flowing from the former to the latter. In NCM, specialized suppliers are on the whole the traditional machinery suppliers that have succeeded in integrating the new electronics technology; whilst in CAD and robots, they are on the whole new entrants spinning off from large-scale users. In a thorough survey, the OECD has recently shown a similar pattern in automation-related software, with significant contributions being made by both specialized suppliers and large-scale users.[19] Many of the latter have established their own automation centres, and some of them are diversifying vertically into automation engineering.

As Rosenberg has shown,[20] this pattern of interdependent users and producers of capital goods has always been central to technological change in the production-intensive sectors. He also identified two other essential features in the pattern, the first of which is the important stimulus to technical change provided by technological imbalances in interdependent production systems. Interdependencies and bottlenecks were particularly important in determining the early rate and direction of development of automation technologies. Numerically controlled machining was essential for aircraft wing configurations proposed in the early 1960s, and CAD was a complement in the design of these configurations. Similarly, CAD/CAM were essential for the design and manufacturing of large-scaled and very large-scaled integrated circuits.

Rosenberg has also stressed the importance of 'vertical disintegration' and 'technological convergence' in the development and diffusion of technologies amongst firms and sectors. However, extension of CAD/CAM to other industries will require the solution of further technical problems. Arnold and Senker[21] have pointed out that the design, machining and assembly of mechanical components that are three-dimensional, irregular and heavy have different and in some ways more difficult technical requirements than those for electronic components that are two-dimensional, regular and light in weight; whilst applications in clothing require, as Hoffman and Rush[22] pointed out, the solution of the long-standing problem of the automated handling of non-rigid materials. Many of these technologial bottlenecks are non-electronic in nature, and often require improvements in mechanical technology: for example, Rendeiro[23] points out that improvements in performance resulting from NCM have depended in part on the finishing and accuracy of lead screws and other mechanical components.

As and when technological bottlenecks have been overcome, the benefit of adopting electronics-based assembly technology have been multiform. Labour saving has remained a major trajectory. Gönenc[24] has suggested—as Rosenberg[25] has done for the nineteenth century—that the stimulus comes not just from increasing labour costs, but from the desire to decrease dependence on an unreliable labour-force. He also suggests that—at least in Scandinavia—it has come from the pressure to eliminate labour from tedious and dangerous work. However, most analysts identify another set of benefits which they give at least equal importance. These are improved product performance and quality, increased speed of product development, reduced holdings of inventory, and increased flexibility in production that can result from the combined uses of CAD, NCM and robots.[26]

Supplier-dominated firms

The channels of diffusion of electronics-based production into the supplier dominated sectors in traditional manufacturing are similar to those of vertical disintegration and technological convergence described by Rosenberg.[27] The development and use of the technology in the leading user sectors (i.e. aerospace, automobiles, electrical/electronics) improves performance and reduces costs to a level where specialist firms staffed with personnel previously trained in these user firms can begin to adapt and simplify the technologies for use in supplier-dominated firms. Hoffman and Rush[28] have described and analysed in some detail how this process has affected the clothing industry. CAD has begun to revolutionize the design process, and both NC sewing machines and robots are to be used in the industry. Most of these innovations have come from both traditional and new equipment suppliers, although a few large user firms in clothing have also made a contribution, and large customers of the clothing industry continue to put pressure on the many small and technically understaffed firms in the industry to adopt the latest technology.

In other words, the sources of technical innovation in clothing remain largely outside the industry, even though the nature and the pace of such innovation are being profoundly modified. Whilst this is likely to remain true for most of traditional manufacturing and for many services, a significant shift in the sources of innovation is under way in certain firms with heavy and complex operations in information processing. Decreasing costs and increased performance of equipment have opened new opportunities for increasingly automated and interdependent systems of information processing, progressively replacing the free-standing information processing 'tools' (typewriters, computers, filing systems) that had existed previously. These organizations have developed professionalized 'systems engineering' groups, whose functions are similar to 'production engineering' and 'process engineering' groups in scale-intensive manufacturing, namely, to specify, assimilate and operate increasingly complex and interdependent 'production' systems. Over time, these user organizations have developed the capacity to write their own software; to modify, design and specify their own interdependent systems; and eventually to build their own hardware. As a result, the sources of technical change move progressively from complete reliance on general purpose equipment suppliers to an increasing contribution by large-scale user firms themselves, and by specialized suppliers of hardware and software, many of which are 'spin-offs' from users.

The recent OECD report confirms this trend in data processing software.[23] Over time, the share of production has shifted markedly from computer suppliers to computer users, and to newly formed software houses. The study by Barras and Swann of the British insurance industry shows the same trend, and explains that it results from incompatibilities amongst bits of equipment, from lack of appropriate software from general purpose suppliers, and from product differentiation.[29] Both studies confirm the importance of specialized data-processing groups in user firms.

However, some of the findings suggest that this trajectory of technological development may not continue. Trends in hardware and software are inevitably dominated by the wide range of products offered by IBM.

Deviation by users from their range as a result of in-house development generally leads to big adjustment costs back to the dominant (i.e. IBM) trajectory. Specialized data-processing groups in user firms can be harmful to the extent that they reduce the degree to which the design of products and production systems are compatible with package software provided by suppliers; and the role of such groups may diminish with the advent of distributed processing. Equipment supplier firms, on the other hand, may reverse the trend and increase their contribution to software production, as a result of technical change in hardware and software, and of changes in their strategies.

The semiconductor and the size structure of innovating firms

We would expect that the size structure of innovating firms will reflect both earlier technological trajectories, and the specific effects of semiconductor technology. In the electronic/electrical sector, successfully innovative firms will continue to grow through product market diversification. Those that have successfully exploited semiconductor technology since the 1950s have already become very big. Freeman[30] pointed out that employment in IBM increased from 22,000 in 1946 to 205,000 in 1971, and by now it is more than 350,000. However, experience over the past thirty years shows that these large organizations do not succeed in exploiting commercially all of the considerable number of product market opportunities emerging from their technological activities; some of the 'spin-off' firms started by their former personnel can become very big.

We would also predict that developments in microchip technology will enable the emergence of an increasing number of small, specialized and technologically progressive firms. Some of these—like suppliers of electronic chip-making machinery—will emerge in symbiosis with the mass production of electronic products.[31] In addition, just as the provision of cheap and high-quality steel in the nineteenth century enabled the formation of myriads of small machinery firms, capable of matching steel to a great variety of differentiated mechanical functions, so the provision of cheap information processing capacity in the microchip will enable an increasing number of small firms to match this capacity to an equal variety of differentiated information and control functions. The formation of specialized firms in instrumentation engineering, described earlier, is an early manifestation of this trend. Just as the typical technological activity of the small machinery firm is machine design, so the typical technological activities of the small information-based firms will be software, and the design of special chips and of interfaces with bigger systems. Thus, the semiconductor may result in a bimodal size distribution of innovating firms in the electronics/electrical industry, with both increasingly large, innovative and diversified firms, as well as an increasing number of small and specialized ones.

Some writers have argued that quite the reverse will happen in the production-intensive categories of firms: on the one hand, assembly-based firms can survive being somewhat smaller, whilst machinery suppliers can survive only by getting bigger. In the major assembly-based industry—automobiles—Altshuler *et al.*[32] have documented the adoption of modern and flexible automation. In contrast to the earlier dedicated automation, it can produce and assemble components for a variety of models and variants. It will reduce the volume and increase the skill of labour required, reduce economies of scale, and increase flexibility. The authors argue that medium-sized automobile producers will not in future suffer from cost disadvantages, or from over-commitment to the one model or variant.

At the other extreme, it is often argued that specialized firms supplying capital goods will have to increase in size in the future, given that increasing flexibility will enable standard machines to be manufactured in series or large batches, and then to be programmed for specific uses.[33] It seems to this writer that such a trend is unlikely for two reasons. First, it ignores the enormous *mechanical* diversity of machining requirements. It is unlikely that the same machine could be used efficiently for, say, manufacturing gearboxes and cylinder blocks. What flexible automation allows is the manufacture of more than one type of gearbox on the same machine, or of cylinder block on another machine. Second the argument ignores the opportunities open to the machine builders themselves, as a result of flexible automation: CAD, NCM and robots enable small and specialized supplier firms both to adapt their products to the specific requirement of users, and to take advantage of any economies of scale in component production or purchase. As a result, we would expect small and specialized machinery firms to continue to be able to design and sell equipment specifically adapted to the requirements of the large and sophisticated users.

However, larger firms amongst equipment suppliers might gain another advantage as a result of the application of semiconductor technology. Flexible automation is opening up new markets in small-firm traditional manufacture and services, because it has become relatively cheap, and because it can increasingly be used in small batch or one-off production.[34] These user firms are unlikely themselves to become important sources of new technology, even if automation increases their relative efficiency. Larger equipment-producing firms will have an advantage in these markets, since a strong capability will be necessary for the design and engineering of complete production systems, and the education of users.[35]

Creative destruction or creative accumulation?

To what extent has the pervasive diffusion of semiconductor technology been a process of 'creative destruction;, where the new displaces the old, or of 'creative accumulation', where the new builds on the old? The diffusion of what Freeman and his colleagues have called a 'new technology system'[36] involves—almost by definition—the infusion of new skills, and a speeding-up of the rate of advance of frontier, state-of-the-art technology. These factors in themselves might appear to be generating a process of creative destruction by speeding up the growth and decline of firms. As Nelson and Winter[37] have shown, industrial structures are more volatile when the rate of technical change is faster, and when competitive advances in skills and technology are more easily appropriated by innovating firms. However, these more rapid rates of change of structures and market shares—or what Klein[38] has called 'fast history'—are not synonymous with our definition of 'creative destruction', which requires in addition that growth be associated with new skills and organizations independent from old ones. Our theory of technical change, which stresses the cumulative and differentiated nature of technical change, suggests that accumulation and complementarity of skills in existing organizations typify the diffusions of a new technology system rather than destruction and displacement. Most of the evidence suggests that this is the case.

To begin with, there is the stability within the fast-moving electrical/electronic sector itself. Table 1.5 identifies seventeen American and West European firms as leaders in American patenting activity in key areas of electronics in the 1970s. Ten of these had aready been identified by Freeman (in *The Economics of Industrial Innovation*) as leaders in British patenting in electronic capital goods and office machinery in the 1930s, late 1940s and 1950s: IBM, Philips, RCA, Westinghouse, Bell, General Electric, Siemens, Burroughs, Sperry, Rand and ITT. The remaining seven can be considered as new entrants that have grown very rapidly as a consequence of relatively high rates of innovation: Texas Instruments, Motorola, Honeywell, Hewlett-Packard, Xerox, Data General and Rockwell. Whilst this rate of entry of new firms is probably higher than in most other sectors, it cannot be described as 'creative destruction', since old-established firms continue to create technically at a higher rate than the new entrants. However, the semiconductor has changed the shape of the firms and the competition within the industry, since it has led to growing tehnological convergence and interdependence among office machinery, consumer electronics, telecommunications, and industrial automation.[39]

Stability also typifies the applications of the semiconductor in

Table 1.5 Shares of top ten organizations in American patenting in selected areas of microelectronics since 1969

Patenting organization	Integrated circuit structure patents		Patenting organization	CPUs and other systems patents		Patenting organization	Digital logic circuits patents		Patenting organization	Semi-conductor memories patents		Patenting organization	Speech analysis and synthesis patents	
	%	No.		%	No.		%	No.		%	No.		%	No.
IBM	12.2	289	IBM	12.2	73	IBM	12.6	199	IBM	16.9	194	Bell	12.2	85
Texas Instrs.	8.2	194	Texas Instrs.	12.0	72	RCA	5.9	93	Texas Instrs.	6.7	77	IBM	4.2	29
US Philips	6.9	163	Burroughs	5.7	34	Motorola	5.4	85	RCA	5.7	65	NEC	3.1	23
RCA	6.5	153	Motorola	4.7	28	Bell	4.1	65	Bell	5.7	65	US Navy	2.7	19
Hitachi	5.5	131	Honeywell	4.3	26	Hitachi	3.7	58	Siemens	4.4	50	Texas Instrs.	2.3	16
Motorola	5.1	121	Hewlett-Pack.	2.8	17	Texas instrs.	3.2	51	Gen. Elec.	3.3	38	US Army	1.9	13
Westinghouse	3.4	80	Bell	2.5	15	Rockwell	3.0	47	Motorola	3.2	37	US Philips	1.9	13
Bell	3.3	79	Sperry Rand	2.5	15	Toshiba	2.6	41	Hitachi	3.1	35	Hitachi	1.4	10
Gen. Elec.	3.3	78	Xerox	2.3	14	Siemens	2.5	39	Westinghouse	2.8	32	ITT	1.4	10
Siemens	2.6	61	Data General	2.2	13	Westinghouse	2.1	33	Sperry Rand	2.6	30	Sharp	1.3	9
Top 10 above	57.0	1,349		51.3	307		45.0	711		53.4	623		32.7	227
Total	100	2,367		100	599		100	1,579		100	1,147		100	695

Note: For IC structure and CPUs, organization patents granted 1/1969–12/1979, Total from 1/1969–12/1979. For digital logic circuits, semiconductor memories and speech analysis, organization patents granted 1/1969–6/1982. Total from 1/1969 to 12/1982.

Source: Office of Technology Assessment and Forecast, *Patent Profiles: Microelectronics I*, and II, Washington: US Department of Commerce, 1981, 1983.

production-intensive sectors. Large-scale and established users in sectors typified by continuous processes and mass assembly have played a major role in developing control instrumentation technology and flexible automation. As with earlier generations of capital goods, development and diffusion have happened in symbiosis with small and specialized supplier firms. In the case of NCM, these suppliers are in general the long-established ones who have successfully assimilated electronics technology and skills into their machine design and operation; Rendeiro[40] shows that, in NC machine tools, manufacturers of numerical controls have on the whole remained separate and distinct from the machine builders. Even in the case of CAD and robots, where equipment suppliers have often been new entrants, they have generally emerged from large-scale users.

There is, none the less, case evidence of abrupt and discontinuous changes resulting from microelectronics: for example, watches, meters, calculators, printing and publishing. One possible explanation is that they are products that have a high informational content, and where there has been a radical switch from electromechanical to electronics-based technology. However, if this were a universal law, IBM would have been destroyed in the office machinery market in the 1950s by the established electronics companies. Similarly, established printing machinery companies would have been displaced by electronics-based ones; but Haywood[41] shows that some of them have successfully diversified into the electronics-based technology.

An alternative explanation is that it is not the *type*, but the *level* of technology in established firms that matters. If firms are in sectors that do have a strong technological component, and if they are at the frontier of established technology, they are more likely to be successful and efficient in assimilating the new technologies than firms that are not. This is a stronger hypothesis than one allowing accumulation from one technological regime to the next. It says that being good at the old is a necessary condition for being good at the new. It is consistent with the stabilities observed in electronics and scale-intensive sectors, and with the estimate of Cranfield Institue of Technology that 60 per cent of robotics specialists in the United Kingdom have their previous experience and training in mechanical and production engineering.[42] It is also consistent with Mowery's observations after examining large American firms' R & D expenditures in the period from 1921 to 1946:

> research investment seems to have acted as a form of insurance for the already large firms, reinforcing the position of dominant firms . . . rather than precipitating significant turnover. Such an interpretation is consistent with the evidence . . . concerning reductions in the rate of turnover among the largest American firms since 1920.[43]

Mowery stresses the development of an oligopolistic market structure as the primary explanation of these trends. However, there are at least three other explanations related to the characteristics of technological accumulation.

First, firms accustomed to managing professional scientists and engineers in the old technologies may be more likely to learn how to attract and to use scientists and engineers in the new technologies. In his study of textile machinery, Rothwell[44] found that firms with better developed R & D laboratories were better able to assimilate and exploit advances from non-mechanical technologies, especially aerodynamics, fibres and electronics.

Second, as we have aready pointed out, the full exploitation of rapid progress in electronics technology often depends on upstream or downstream improvements in more conventional technologies. Particularly revealing evidence of this comes from two countries which, if the level of adoption of robots is any guide, are particularly successful in the exploitation of electronics-based technology, namely, Sweden and Japan. Thus, based on a detailed study of a Swedish engineering workshop, Eliasson concludes that the introduction and spread of electronics will be a relatively slow and piecemeal process, given 'unsatisfactory precision and reliability of measurement and sensory equipment and crude mechanical installations that lay behind in development'.[45] Similarly, a recent Japanese survey of the spread of flexible automation identifies '[the] development of high precision in mechanical technology'[45] as one of four major preconditions for its development and adoption. Compared to conventional mass production systems, flexible automation requires greater process accuracy, given more restricted opportunities for learning by doing.

Third, there is the evidence that effective assimilation of electronics-based technology depends on more general competence in management. Eliasson concludes that one of the most important features determining the assimilation of electronics-based technology in Sweden is 'the lack of centralized knowledge of the production process itself'.[47] Arnold and Senker have called this the 'computerisation effect': according to Arnold 'computers impose a need for orderly, clearly defined systems: they are fast but stupid, unlike people who are slow but intelligent enough to muddle their way through ill-defined procedures'.[48]

From such a perspective, Japanese management innovations introduced in mass assembly can be seen as an essential *prerequsite* for the introduction of flexible automation, since they help make explicit the essential features of conventional production technology, as well as increase the incentives for quality control. According to Altshuler *et al.*, the Japanese approach assumes that if production workers are given the skills and responsibilities to diagnose problems, repair equipment and spot defects, then the ranks of supervisors and machine repairmen can be greatly thinned even as quality is improved . . . [and] asserts that high buffers hide problems rather than providing time for their repair; that defect prevention by workers is far superior to defect detection by supervisors; and that ideas for improving the production process can come largely from the line workers who know the system best.[49]

Conclusions

The empirical data on which the above interpretations are based are highly imperfect. They are also incomplete, given that the proposed taxonomy in Table 1.1 does not cover what might be called a 'state capitalist' category of firm, whose products tend to be large-scale and sophisticated capital goods used by organizations whose purchasing decisions are heavily influenced by governments: in particular, those related to defence, energy, transport and communication, the first and last of which are of central importance in the diffusion of semiconductor-based technology. Although the data are tentative and incomplete, their conclusions are clear and straightforward. Sectoral differences in the sources and directions of technical change are similar both before and after the assimilation of semiconductor technology; and the adoption of electronics-based technology is largely a cumulative process, with the new building firmly on the old.

Thus, in electronics we continue to see the growth of large firms out of technology-based diversification into new and growing product markets; in production-intensive firms, symbiotic links remain between large-scale users, and small and specialized suppliers of capital goods for the development of electronics-based production technology; and in supplier-dominated firms, dependence on others continues for applications of semiconductor technology. Compared to existing 'trajectories', the main changes are a more rapid growth of electronics than chemicals firms; the growing importance of small and specialized electronics-based firms providing equipment and software; greater flexibility in scale-intensive firms; and an emerging technological competence in service firms that are large-scale processors of information. A number of other conclusions, both practical and conceptual, also emerge from the analysis.

First, policies to encourage the introduction and diffusion of micro-electronics technology should go beyond support for large firms to include small electronics-based firms, and both large and small-scale users of equipment embodying electronics-based functions. It should also go beyond electronics technologies to include other technologies that are complementary, and potential bottlenecks. According to a British industrialist, R. Curry, writing in 1982:

> I can purchase 'chips' for pennies but I cannot purchase reliable components to interface them to mechanical mechanisms. This means that while it is relatively easy to design calculators and TV games, it is not easy to use microelectronics in the automotive or engineering capital equipment industries. . . . Research into microelectronics without corresponding research into, for example, the more mundane problems of the electronic–mechanical interface means that micro-electronics cannot successfully or easily be used in product design.[50]

Second, given this interdependence between established and radical technologies, it is important to distinguish a 'fundamental technology' (i.e. one enabling the emergence of a new, pervasive technological family) from a 'strategic technology' (i.e one whose control enables a firm or a country to achieve competitive success). Thus, whilst the semiconductor is certainly a fundamental technology, it is not necessarily strategic, but will influence strongly the strategic opportunities that do emerge. For example, firms at the frontier of flexible automation do not necessarily control the relevant semiconductor technology itself, but the interface between it and conventional mechanical and production engineering.

Third, it is unrealistic to expect that—as has been suggested by certain analysts[51]—developing and newly industrializing countries will be able to 'leapfrog' mature industrial countries in electronics technologies, because of the latters' strong commitment and competence in older and slower-moving technologies. Where—as is often the case—there are strong complementarities between old and new technologies, developing and newly industrializing countries will be constrained to follow similar technological trajectories to mature industrialized countries. Countries like Japan (and earlier Germany) that have caught up and overtaken leading countries have done so through rapid technological accumulation in both old and new technologies.

Fourth, the exploitation of even radical technological change—whether at the level of the firm, sector or country—is unlikely to be random but to build upon, or grow out of, previously existing competence. Thus, the rapid rate of development and adoption of robots in Swedish industry[52] reflects in part a high pre-existing level of technological competence in mechanical and production engineering. For similar reasons, we would predict that firms with strong technological competence in pharmaceutical products are likely to become strong in future in the exploitation of biotechnology.

Finally, no simple model is likely to describe and explain satisfactorily the sources and directions of technical change in the emergence and diffusion of a major new technology. Thus, the 'product cycle' model,[53] where fast-moving technologies emerge through new small firms and product innovation and then gradually stabilize technologically into large firms concentrating on process innovations, is inadequate. Although new small firms have played an important role in the development of electronics technology, major contributions have been made from the beginning by large firms—in electronics, office machinery and automobiles—and by their spin-off suppliers. Furthermore, a series of *process* innovations in the manufacture of semiconductor devices have been fundamental from the beginning, and have been the basis of subsequent product innovations.

Similarly, explanations of the rate and direction of technical change based entirely on economic signals and institutional constraints have limited explanatory power. For example, they neglect accumulated mechanical and production engineering skills in explaining international variations in the adoption of robots. Or they are tempted to explain the advent of flexible automation as a rapid response in the 1970s to the requirements of slow and volatile economic growth, and to neglect the contribution of the semiconductor, and of twenty years of accumulated experiment and learning in CAD, NCM and robots.

The concepts of technological 'trajectories' or 'paradigms' as developed by Nelson and Winter, Rosenberg, Dosi[54] and this writer may be a more promising analytical path to follow, since they recognize the cumulative and differentiated nature of technology. They also allow that the logic of technology itself does strongly influence—if not determine—the rate and direction of technical change.

Notes

I wish to thank the following colleagues for helpful suggestions and criticisms on an earlier draft of this paper: J. Gershuny, K. Hoffman, D. Jones, M. McLean, J.-J. Salomon, P. Senker.

1. S. Cole, C. Freeman, M. Jahoda and K. Pavitt, *Thinking about the Future: A Critique of 'The Limits to Growth'*, London, Chatto and Windus, 1973.
2. C. Freeman and M. Jahoda, *World Futures: The Great Debate*, London, Martin Robertson, 1978.
3. C. Freeman, J. Clark and L. Soete, *Unemployment and Technical Innovation: A Study of Long Waves and Economic Development*, London, Frances Pinter, 1982.
4. C. Freeman, *The Economics of Industrial Innovation*, London, Frances Pinter, 2nd edn., 1982.
5. J. Townsend *et al.*, *Innovations in Britain since 1945*, SPRU Occasional Paper No. 16, University of Sussex, 1981.

6. K. Pavitt, 'Sectoral Patterns of Technical Change: Towards a Taxonomy and a Theory', *Research Policy*, **13** (1984), pp. 343–73.
7. J. Kamin *et al.*, 'Some Determinants of Cost Distributions in the Process of Technological Innovation', *Research Policy*, **11** (1982), pp. 83–94.
8. K. Pavitt and L. Soete, 'International Differences in Economic Growth and the International Location of Innovation', in H. Giersch, (ed.), *Emerging Technologies: Consequences for Economic Growth, Structural Change, and Employment*, Tübingen, J. C. B. Mohr, 1982; D. Teece, *The Multinational Corporation and the Resource Cost of International Technology Transfer*, Cambridge, Mass., Ballinger, 1977.
9. K. Pavitt, 'Some Characteristics of Innovative Activities in British Industry', *Omega*, **11** (1983), pp. 113–30.
10. R. Nelson and S. Winter, *An Evolutionary Theory of Economic Change*, Cambridge, Mass., Harvard University Press, 1982.
11. Townsend, *et al.*, op. cit. note 5.
12. C. Perez, 'Structural Change and the Assimilation of New Technologies in the Economic and Social Systems', *Futures*, **15** (1983), pp. 357–75.
13. J. McLean and H. Rush, *The Impact of Microelectronics in the UK: A Suggested Classification and Illustrative Case Studies*, SPRU Occasional Paper No. 6, University of Sussex, 1978.
14. J. Northcott, P. Rogers and A. Zeilinger, *Microelectronics in Industry: Survey Statistics*, London, Policy Studies Institute, 1982. A similar and more recent survey in 1983 confirms these results: J. Northcott and P. Rogers, *Microelectronics in British Industry: The Pattern of Change*, London, PSI, 1984.
15. E. Arnold, *Computer-Aided Design in Europe*, Sussex European Papers No. 14, University of Sussex European Research Centre, 1984.
16. R. Gönenc, *Électronisation et Ré-organisations Verticales dans l'Industrie*, Unpublished Thèse de Troisième Cycle, University of Paris X, Nanterre, 1984.
17. B. Carlsson, 'The Machine Tool Industry—Problems and Prospects in an International Perspective', *The Industrial Institute for Economic and Social Research*, Stockholm, December 1983, Working Paper No. 96.
18. 'Preconditions for Flexible Manufacturing System', *Mechatronics News*, **1**, No. 1 (1983), pp. 3–4.
19. OECD, *Software: A New Industry*, Committee for Information, Computer and Communications Policy, document ICP (84)4, February, 1984. I am grateful to Mr H. P. Gassman for bringing this excellent synthesis to my attention. The main contributor to its preparation was R. Gönenc.
20. N. Rosenberg, *Perspectives on Technology*, Cambridge, Cambridge University Press, 1976; N. Rosenberg, *Inside the Black Box: Technology and Economics*, Cambridge, Cambridge University Press, 1983.
21. E. Arnold and P. Senker, *Designing the Future: The Implications of CAD Interactive Graphics for Employment and Skills in the British Engineering Industry*, Engineering Industry Training Board, Occasional Paper No. 9, 1982.
22. K. Hoffman and H. Rush, 'Microelectronics and Clothing: The Impact of Technical Change on a Global Industry', Science Policy Research Unit, University of Sussex, mimeo. 1983.
23. J. Rendeiro, *Technical Change and Strategic Evolution in the Machine Tool Industry*, Sussex European Research Centre, mimeo. 1984.
24. Gönenc, op. cit., note 19.
25. Rosenberg, op. cit., note 20.

26. P. Senker, 'Some Problems in Implementing Computer-Aided Engineering—A General Review', *Computer Aided Engineering Journal*, **1** (1983), pp. 25–31.
27. Rosenberg op. cit., note 20.
28. Hoffman and Rush, op. cit., note 22.
29. R. Barras and J. Swann, *The Adoption and Impact of Information Technology in the UK Insurance Industry*, London, Technical Change Centre, 1982.
30. Freeman, op. cit., note 4.
31. M. McLean, 'Chip Makers Report Economic Lift', *Electronics Times*, 2 June 1983, p. 6.
32. A. Altshuler, M. Anderson, D. Jones, D. Roos and J. Womack, *The Future of the Automobile: Report of MIT's International Automobile Programme*, London, Allen and Unwin, 1984.
33. Gönenc, op. cit., note 19.
34. T. Sasaki, 'Microcomputers in the Japanese Consumer Durables Industry—Status and Prospects', in M. McLean (ed.), *The Japanese Electronics Challenge*, London, Frances Pinter, 1982; Carlsson, op. cit., note 17.
35. Carlsson, ibid.
36. Freeman *et al.*, op. cit., note 3.
38. B. Klein, *Dynamic Economics*, Cambridge, Mass., Harvard University Press 1977.
39. Sasaki, op. cit., note 33.
40. Rendeiro, op. cit., note 23.
41. B. Haywood, *Technical Change and Employment in the British Printing Industry*, World Employment Programme Research, International Labour Organisation, Geneva, September, 1982.
42. Gönenc, op. cit., note 19, Figure 26.
43. D. Mowery,*The Emergence and Growth of Industrial Research in American Manufacturing*, Unpublished Ph.D. dissertation, Stanford University, 1980, p. 7.
44. R. Rothwell, *Innovation in Textile Machinery: Some Significant Factors in Success and Failure*, SPRU Occasional Paper No. 2, University of Sussex, 1976.
45. G. Eliasson, 'Electronics, Economic Growth and Employment—Revolution or Evolution', in Giersch, op. cit., note 8, pp. 77–95.
46. Op. cit., note 18, p. 1.7.
47. Arnold and Senker, op. cit., note 21.
48. Ibid., p. 6.
49. Altshuler *et al.*, op. cit., note 31.
50. R. Curry, Written submission to the House of Lords Select Committee on Science and Technology, *Sub Committee on Engineering Research and Development*, Vol. III—Written Evidence, London, HMSO, 1983, pp. 75–7.
51. L. Soete, 'International Diffusion of Technology, Industrial Development and Technological Leapfrogging', *World Development*, **13** (1985), pp. 409–22.
52. 'Robots: The Users and the Makers', *OECD Observer*, No. 123 (July 1983), pp. 11–17.
53. R. Vernon, 'International Investment and International Trade in the Product Cycle', *Quarterly Journal of Economics*, **80** (1966), pp. 190–207; J. Utterback

and W. Abernathy, 'A Dynamic Model of Product and Process Innovation', *Omega*, **3** (1975), pp. 639–56.

54. G. Dosi, 'Technological Paradigms and Technological Trajectories', *Research Policy*, **11** (1982), pp. 147–62.

2 Innovation in materials

George F. Ray

Introduction

This paper starts with the proposition that without scientific and technological advance—in other words, without innovation—mankind would long ago have had to face severe material shortages. In the past, much public anxiety has been expressed about the consequences of such shortages,[1] but new materials have been discovered and new methods introduced to produce and process them. Plastics and other synthetic materials, discussed by Freeman,[2] are cases in point.

At the time of writing, in 1985, almost all materials, both conventional and relatively 'new', are available at historically rather low real prices in virtually any quantity. Although the world's non-renewable resources are obviously finite, the lifetime of the known and proven reserves is, in general, reassuringly long.[3] Hence, one might ask, why bother? While the question seems legitimate, the problem is more complex than appears at first sight; the whole truth needs to be approached from several angles.

First, while overall physical scarcity of industrial materials is not a likely eventuality—as an OECD study put it[4]—the real problem is undisturbed access to them. World reserves data conceal the fact that the reserves and production of more than a dozen important minerals/metals are highly concentrated in a small number of countries (such as South Africa, Brazil, Zaïre, Chile, Morocco, China and the USSR) and any disturbance in production or transportation in any of these countries can lead to temporary shortages.

Second, there may be a sudden demand for a particular material stemming from some novel end-use. For example, the price of rhodium, one of the platinum-group metals, trebled in the twelve months to March 1985, despite the weak metal market: this was due to a demand from the automobile industry which required rhodium for its own technological development, aimed at reducing the damaging effects of exhaust fumes.

Third, in the summer of 1983, the UK Department for Industry set up a Materials Advisory Group (now known as the Collyear Committee) whose work resulted in the publication early in 1985 of 'A programme for the wider application of new and improved materials and processes'.[5] This report stresses the outstanding importance of such work, strongly advocating the need for government involvement in the research, development and application of new materials. Thus, there is plenty of evidence indicating the topicality of the subject.

Another question that may be raised is this: what has been the contribution of basically economic studies, of analyses by economists, to progress in the materials area? Naturally, the lion's share in this sphere, as elsewhere in technical progress, goes to scientists and technologists. Nevertheless, the work of economists and historians should not be belittled: they greatly contributed to the understanding of the process of innovation in materials and the forces at work,[6] to the organization of research and development,[7] to the search for the direction of desirable research,[8] to various aspects of the economics of novel materials,[9] and to being generally better informed, to foresee and anticipate troubles. As a historian wrote: 'it is surely arguable that foresight and anticipation may assist in preparing alternatives and make more likely a successful response to critical shortages'.[10]

The scope of this paper

This study is in two main parts. The first surveys the history of some thirty materials which are relatively new or were 'new' at the time of their introduction into general use. The dissemination of any innovation takes time; that of new materials often takes an unusually long time; hence this first part concerns materials that already have a history. The second, shorter, part deals with the present, that is, the progress of the area of materials. For both the past and the present the choice is very wide and therefore it was necessary to be selective.

Many of the 'new' materials were discovered or developed as the outcome of a specific need under wartime or market pressure. Others were the result of spontaneous and random scientific/technological advance. The various materials are thus crudely categorized, although these definitions are frequently blurred and inevitably overlap. Because of limitations on the length of this contribution, detailed case histories of the various materials cannot be included; they may be found elsewhere.[11]

The role of science and technology, as applied to industrial materials and their use, is complex, extending well beyond the relatively simple concept of 'new' materials. We therefore start by discussing briefly some of the complexities involved.

The impact of technology

Although important, the discovery or development of a new material is still only one of the ways in which supplies may be affected through scientific or technological progress. The supply system is complicated by interaction between miners and farmers through traders, exchanges and middle men, designers, processors and consumers to recycling. In principle, it is subject to market forces at every single stage and in practice it is further coloured by state intervention. Any national system has its roots in a world market and this gives it a truly international flavour. The technological side is equally complex: it affects not only all aspects of the material itself, from exploration to processing, but also its application in end-use.

Originally, there was only neutral matter in nature. A material became a natural resource when man began to use it for a specific purpose. Coal was a useless black rock before man started to burn it; bauxite did not even have a name before it was discovered that it could be processed into aluminium. The promotion from neutral matter to resource was itself an advance. At all later stages technology and science play significant parts and their spectrum is incredibly wide: 'materials technology' (narrowly defined) makes use—particularly nowadays, in view of very rapid communications—of interventions and innovations in almost any area or discipline. Aerial photography is commonly used to assist in the exploration for mineral deposits, analysis of local flora for the existence of metals in the subsoil and techniques developed in medicine may be used in materials production (such as that used in the prevention of blood clotting applied to rubber trees to increase yield), to mention just a few of the many examples of technological cross-fertilization.[12] The vast development of mining technology is well known, including methods of strip mining, a technique introduced after, and with the experience gained from, the digging of the Panama Canal.[13]

Technology has helped to reduce production costs and this has been of fundamental importance, for example in copper mining. In the early nineteenth century, Cornish copper ore, with a metal content of 13 per cent, was the price leader; by the middle of the century, after the exhaustion of Cornish mines, Anglesey copper took over, with a metal content of about 3 per cent. New mines nowadays contain ores with around 0.5 per cent metal, and yet the real price of copper has decreased very considerably over time.[14] It is technological advance—and by this we mean a long chain of major and minor innovations—that has made cost-efficient production possible, using leaner ores and reducing the cost and price of the final primary metal still further.[15]

Another advance, the Thomas process, brought high-phosphoric ores into the ambit of valuable steelmaking resources. They had previously been useless. Thus, for example, the French were able to use highly phosphorous Lorraine ores for steelmaking from about 1880.[16] Similarly, around the middle of the present century, many of the major American steelworks were literally saved from extinction by the introduction of pelletized taconite as their base material, replacing the exhausted high-grade conventional iron ore from the Mesabi Range of Minnesota.[17] Both these cases represent materials that were 'new' at the time. There are many other examples of technological 'intervention' in conventional methods of material production and use.

Lead times

It is commonly held that modern communication speeds up everything and the spread of innovations is no exception. There is plenty of scope for speeding up in the field of new materials, as can be seen from the following few historical examples.

Coal was used by the Romans more than 1,600 years ago; monastic records reveal coal-extracting operations in the twelfth century in Britain and Germany. Yet the widespread use of coal had to wait for two major innovations in the eighteenth century, the steam engine and the replacement of charcoal with coal in ironmaking, after which coal became acceptable to industry.[18]

The *oil* produced by Drake's 1859 well or that from the Baku oilfield in Romania, was first used as a medicine or for lighting; the real take-off for oil began much later, in the present century.

Indian *cotton* became familiar to the Greeks through the advance of Alexander the Great (323 BC); the Romans produced inferior quality cotton in Malta, but cotton fabrics remained a luxury until the end of the eighteenth century, when Britain's North American colonies started to cultivate it. The trade then took off rapidly, aided by a series of major innovations at all stages.[19]

After first having been isolated in 1825, *aluminium* was for a long time considered a precious metal: 'silver from clay', as exhibited at the 1855 Paris Exposition, used for cutlery at court banquets, a rattle for the Prince Imperial at the time of the Second Empire in France, or a watch charm for the King of Siam. Its electrolytic reduction was developed in 1886, but processors did not know what to do with it, and wider applications had to wait until the present century.[20]

Tungsten has been known to science since the end of the eighteenth century but its use in industry did not begin until around 1900.[21]

Thus, the time period from invention or discovery of a new material and its innovation to its wider diffusion, i.e. commercial production and adoption by users, can be very long indeed. There are several reasons for this: the most important is probably that a 'new' material is often—as Rosenberg formulated—'an example of an invention which could be produced under experimental conditions for many years before means of producing it commercially were developed'.[22]

The effects of war

The textbook example of the pressure on materials during wartime is that of the introduction on the continent of *sugar-beet* during the Napoleonic wars, when France was cut off from her traditional cane sugar supplies by the British fleet.[16] Similarly, at the same time—and perhaps more appropriate to the discussion of industrial materials—the production of *alkalis* (mainly soda) was revolutionized. Widely used, from the making of textiles to gunpowder, *soda* (sodium carbonate) was originally obtained from the ashes of certain plants, grown mainly in Spain and the Canaries. Leblanc, however, worked out a method of making it from common salt, and in 1808, when the French were cut off from their Spanish sources during the Peninsular War, the manufacture of soda by this process began, remaining the main method in use throughout the nineteenth century (Britain took it over in 1823).[16]

Germany's isolation during the 1914–18 War necessitated the urgent large-scale introduction of the *nitrogen* fixation process, developed by Haber a few years earlier. Nitrogen was widely used in various forms, among other things as an explosive, and its traditional source was Chilean nitrate. By the end of the war one half of Germany's huge production of nitrogen compounds was based on the new process. (One cannot avoid some philosophical thoughts in this context, since without this contribution the war would undoubtedly have ended earlier.)[23] Another indispensable base for the making of explosives was *glycerine*; its traditional manufacture required large quantities of fat—very scarce during the Great War; a substitute was developed—again, in Germany—obtained by the fermentation of sugar residues.[13]

The shortage of *fats*—and hence of soap and candles—began to affect Europe around the middle of the nineteenth century when the supply of solid animal facts was unable to keep pace with growing requirements. Liquid vegetable oils were used as a supplement in households but this was not possible in industry. The process of hardening liquid fats by hydrogenation solved the problem. But during the Great War even vegetable oils (and hence soap) became scarce in Germany and this shortage led to the development of the first synthetic *detergent*.[24] It was an inferior product, a true 'Ersatz', yet it deserves attention because its later development was characteristic of all synthetics. Once the basic problem of synthetic production was solved, progress continued. In the following decades, the simple Ersatz was developed into a sophisticated family of products, a wide range of specialized varieties which could serve particular purposes better than the original 'all-purpose' ancestor. This was the case with synthetic detergents, fibres, dyes and rubber alike.

Rubber was perhaps the most important instance of technological development during the 1939–45 War. Hitler's Germany and the United States were equally cut off from Malayan natural rubber. The Germans developed '*buna*', a fairly successful substitute that covered almost all their requirements. Before the war, Du Pont research in the United States had already produced synthetic elastomers with specific qualities, as had Esso for inner tubes of tyres; these innovations proved decisive when the Japanese invasion of South-East Asia compelled the United States to look for another source. The output of all the initial 'new' rubber was still experimental in 1943; only four years later, however, production had reached 100,000 tons and developed into a major industry. This illustrates the rapidity of technological response when the pressure is really great and the backing equally powerful.[13, 25]

Another noteworthy example of progress is that of *acetone*, used as an agent to incorporate gun cotton into nitroglycerine in explosives. Temperate hardwood provided the original source but this became scarce during the Great War; the Germans used various chemicals to replace it, whilst Weizmann in the United Kingdom discovered the process whereby ordinary grain ferments bacteriologically into acetone.[13, 26]

Peacetime scarcities

Wars are obvious causes of scarcity, either because normal trade flows become disrupted by consuming countries being boycotted by their enemies, or by disturbances to the production or transportation of materials. Shortages may, however, also be generated in peacetime, through requirements widely outstripping supplies or through the manipulation of the market by agents with vested interests in achieving monopolistic gains.

Paper was originally made from fibrous textile residues such as rags, but in the nineteenth century the supply of rags was insufficient to satisfy the rising demand from paper mills. In 1841 the Halifax mill was the first to produce 'groundwood paper', using a German patent. Within ten years, the chemical treatment of *woodpulp* was introduced, making it the single base material without which the enormous quantities of paper could never have been produced.[13, 16] Even so, it had to be supplemented by recycling, a process that has also gone through a chain of minor innovations, such as, for example, de-inking.[13]

Material shortages can be artificially engineered, but attempts at cornering the market may seriously misfire. This happened around 1830, when a French syndicate tried to exploit what was then a monopoly of Sicilian *sulphur* the base material of sulphuric acid. Their speculation was overdone: it led to the invention of the process which uses sulphide ores (mainly pyrites).[13,16]

A somewhat similar case is that of *camphor*, which was mainly produced in Formosa. Having occupied that island in 1895, the Japanese attempted to benefit from the steeply rising price of camphor, the production of which had been reduced by wartime conditions; in so doing they stimulated research activity and soon a synthetic substitute was introduced which quickly replaced the natural product in industrial use.[13,28]

The camphor incident, however, had much wider consequences, which are not only important in themselves but also present an interesting case of serendipity (a phenomenon not unusual to scientific researchers). Among others, the Belgian scientist Baekeland (working in America) was unsuccessful in his search for a camphor substitute. While he did not entirely succeed, he did make a systematic study of certain chemical reactions and rather unexpectedly developed a hard, chemically-resistant plastic material which has come to be known as *bakelite.*[29] This was important in its own right but also opened the door to significant further developments, leading eventually to today's enormous family of plastic materials.

Another outstanding factor contributing to the foundations of the later great polymer and pharmaceutical industries was the invention of synthetic *dyes*. There was no shortage of natural dyes but by the middle of the nineteenth century it must have been fairly clear that requirements might soon outstrip the supplies then available. Research began in both the United Kingdom and Germany and in 1856 the German patent application (by BASF) beat the British one (Perkin) by one day! The product was the first synthetic alizarine dye and it paved the way for many other synthetic dyestuffs.[30]

Cryolite is essential in aluminium production as a flux in electrolysis. The largest deposit of it, in Greenland, was totally depleted by 1963, but somewhat earlier a perfect synthetic substitute had been developed.[31] Thus, the advance of technology and science sometimes solves scarcity problems even before they become acute. Synthetic polycristalline *diamonds* perfectly replace the rare natural stone in industrial application;[32] man-made radio-isotopes are successfully substituted for *radium*, the great discovery of the Curies; and, to quote a more recent example, synthetic *quartz* crystals are the accepted substitute for the natural quartz crystal required in a particular quality and in steeply rising quantities by the electronics industry.[33]

There are often spin-off discoveries from research too: it was in the search for a diamond substitute that *tungsten carbide* was developed; this extremely tough material is in many respects superior to the mainly natural materials previously used in the making of dies, cutting tools and so on.[24]

The science 'push'

While the brief case studies in the previous sections deal with new materials that were invented or discovered under some kind of pressure, either from war or from market needs (that is, various forms of 'demand pull'), there are also examples of new departures arising outside such situations, purely as a result of research. One of the best examples is that of *man-made fibres.* At the time of their initial development there was no particular shortage of the different varieties of natural fibres. Although the basic invention goes further back, commercial development started early in this century, with 'artificial silk'—later to be called rayon—followed by nylon in the 1930s and by today's large group of synthetic fibres, including even 'elastomeric' varieties from the late 1950s.[13] Rayon was based on cellulose, the further development of which yielded *cellophane*, a thin, non-fibrous film widely used for wrapping—originally a French product (1912) which was improved by waterproofing in the United States (1920).[30]

Numerous man-made materials that have come to be called by the blanket name *plastics* have, of course, become part of modern life. Specific properties make their wide range—from PVC through polystyrene to transparent polymers—popular in industry and elsewhere, for an endless variety of purposes.[30] In many cases the role of science has been not so much to discover new materials as to find uses for them. *Titanium*—like tungsten, already mentioned—has long been known, but its application remained a problem until a process for refining it was developed in 1936, and its high strength/weight ratio was recognized by the aircraft industry (it accounts for 9 per cent of the airframe structure weight of the jumbo Boeing 747).[24] The first metal in the '*rare earths*' group was discovered in 1794 and in recent decades many members of this category have been used increasingly in industry.[13, 27, 33]

Glass fibre is a new material that has found rapid acceptance on a wide front, from insulation to boatbuilding, and its highly refined version, optical glass fibre, has started to take over the role of copper and other wires in communications systems.[13, 27]

The spectacular advances in the chemical industry outside the field of industrial materials proper are also worth noting. Various types of *fertilizers* have greatly helped to raise food production.[16] Research is well advanced with the objective of producing synthetic *proteins* for feeding animals. But it is in pharmaceuticals that scientific progress has been the most dramatic.

Napoleon's troops used to chew the bark of the willow tree to relieve pain: this bark contained a chemical, related to *aspirin*, which Bayer in Germany started manufacturing at the end of the last century.[13, 29] Many people recall how *penicillin* was also discovered almost by chance.[24] More important, however, penicillin was the first antibiotic, and its later semi-synthetic version was a vast improvement on the first 'natural' variety, and from those early beginnings many types of antibiotics have been developed, many of them serving very specific purposes.

A historical study of the cases listed in this section leaves one with the impression that they were the result of a genuine 'science push', but this elusive concept is difficult to define and delineate or to demonstrate in isolation. Let us take man-made fibres. As already stated, at the time of their introduction there was no particular shortage of natural fibres so that the new material could be treated as the result of science push, a considerable intellectual achievement by those scientists working in basic research. But one can argue just as well otherwise: the key R & D work was done within the research establishments of the large leading enterprises, such as ICI, Du Pont and the German chemical giants; one can assume that a general idea, at least, of the potential of the hoped-for result of these particular projects (if successful) was 'in the air' or perhaps more concretely formulated.

Advocates of 'science push' would reply to this that, while this may have been so, in many other cases no further development could have taken place without the incentive—that most important first step—provided by science. In fact, while it may be of interest to theorize about market pull and science push, both theories seem so closely interconnected in modern life that one cannot but agree with Freeman that, 'it is quite possible to subscribe, at least partially, to both theories'.[34] (See appendix.)

Several serious scholars agree with Freeman, even if they formulate their views differently. Rosenberg says that 'potential demand may exist for almost anything under the sun, and the mere fact that an innovation finds a market can scarcely be used as evidence of the undisputed primacy of 'potential demand pull' in explaining innovation'.[35]

Utterback approaches the question in another way: while writing that 'any innovation is necessarily a combination of a user's need and a technological means to meet that need', he takes the view that 'need input often takes precedence over the technology input' and this is because 'basic research enters the innovation process through education and development of trained people, who later apply in meeting users' needs what they learned years earlier'.[36] However, even for him this is only 'often' the case, leaving Freeman's above-cited thesis still valid.

Some new developments

All the above cases have a history but progress continues and there are many new materials in the R & D pipeline. Some of them are well advanced and have already been used commercially, others are still the well-guarded secrets of their developers. Many of them face the same difficulty that hindered the rapid dissemination of the use of aluminium.

With man-made materials, widespread use depends on a number of factors, of which the most important are availability and price, properties and ease of processing. The latter is of supreme importance; without it the circle can be vicious. For example, carbon fibres encased in plastics could replace many metals (so far as properties are concerned), but processing is slow; therefore they are not available in quantity, hence the price is high and subsequently they may not in general be a commercial proposition. This vicious circle may take some time to break.

There are composite materials that could replace steel or aluminium in the construction of motor vehicles but their processing methods are not to hand and therefore the change does not make commercial sense. All this refers to the long lead times of any new material. The introduction of a new material more often than not requires innovative design and may necessitate redesigning the whole product or the production system.[37] This is risky and expensive: vested interests such as existing capital investment and expertise present constraints in any case so that in order to balance all these handicaps new materials or allied processes are expected to offer substantial cost savings.

Despite these intitial difficulties many new materials have already been commercially applied—if not generally, at least by pioneering companies. *Carbon fibres* are a case in point. Almost parallel development work in the United States, Japan and the United Kingdom was based on the recognition that the carbon–carbon bond is the strongest of any in three dimensions. Interestingly, carbon fibres were first commercially applied in the making of sports goods (golf shafts, fishing rods, tennis racquets); only later—and rather slowly—did the new material start to penetrate the aerospace and other industries.[38]

An important group of new materials is *engineering ceramics*. These are formed from compounds, they are light, stiff, corrosion- and wear-resistant, with low thermal and electrical conductivity—properties that are increasingly attractive in engineering applications.[38]

As one example, we mention *sialon*, based on the alloys of the elements Si, Al, O & N, which has captured a sector of the machine tool tip market and has potential for penetrating other areas.[38] The discovery of 'ceramic alloying' is an example of a generic invention which may have wide applications in many fields. Ceramics are just one segment of *composite materials*—engineering materials built up from several components with different properties (the term as commonly used generally assumes reinforced materials). It is precisely this combination of different properties that represents a new material/technological approach to product manufacture, offering scope for design innovation.[5]

Thermoplastic materials are also relatively new. They are a considerable improvement on the first-generation plastics or polymer matrices that, once set, cannot be reshaped plastically by heating: thermoplastic matrices are amenable to intermediate forms because they can be heated and pressed into moulds, into sheet, tube or any other complex form. This ability rapidly offered new opportunities to the mass production of plastic articles, to highly 'oriented' polymers. Similar experiments with *metal matrix* composites are aimed at reducing the old danger of metal fatigue.[5]

The dramatic progress in the microelectronics industry owes much to developments in materials and their processing. We are probably far from the end of the development work in electronics, and particularly in the materials field. Parallel with work on the engineering application of the results, research is going on into *superlattices* (thin layers of semiconductor materials) whose properties promise to offer opportunities for novel devices and for tailoring the properties of material over a far wider range than is now possible. Other likely departures include: device processing for special uses with materials substantially different from silicon, new chemicals—including organic semiconductors—and so on.[5]

This list of relatively new departures affecting materials could be made much longer. To conclude our tentative illustration of the continuing work in the area under discussion, it should suffice to mention some of the new manufacturing *processes* that will certainly have an impact on the search for materials, probably including new ones, that suit them best. Among them, as examples, it is worth bearing in mind:

– powder metallurgy, which is supposed to avoid the time-consuming and wasteful conventional shaping of parts and components;
– precision casting, which has become imperative in modern aerospace and other industries;
– superplastic forming and diffusion bonding, which reduce several operations into one single process;
– CAD/CAM techniques which will certainly spread from the most complex to simpler manufacturing processes and, particularly when combined with robotics, raise specific requirements of material properties;
– coating technology, which has already reached advanced stages in the fight against corrosion and other harmful effects but is presumed to go much further in surface treatment technology;
– the application of lasers to welding and other forms of joining materials, etc.

Concluding thoughts

In one of his many publications concerning the role of science, Freeman wrote: 'Much scientific research is concerned with the exploration of the unknown. By definition we cannot know the outcome of such explorations and still less can we know its future impact on technology.'[39] This general statement can be applied with some force to materials. This brief survey is intended to show how science and technology have contributed to the supply of industrial and other materials in the past and that progress has been a continuing one. History does not necessarily repeat itself; nor do the examples given in this paper of scientific and technological achievements solving materials problems provide any guarantee for the future. They do, however, provide a basis for the hope that progress will go on and future advance continue to secure a link between the demand for and the supply of materials that industry and other sectors of the economy will require.

Appendix: demand pull or science push?

While we may theorize about these important stimuli to invention it should not be forgotten that much of the equipment, and even food, that has become indispensable in modern life was invented at various times as the result of personal eccentricity or inconvenience. Inventiveness no doubt, but how should one classify it?

The ballpoint pen, nicknamed the 'biro', was devised in 1938 by a Hungarian journalist Laszlo Biro, because he got bored with always blotting his work.[40] The first carpet sweeper was put together by Melville Bissell, a china shop owner in Michigan, because he suffered from headaches caused by an allergy to the dusty straw used for packing his wares.[40] The first zip fastener was demonstrated in 1893 at the Chicago World Fair but it came apart easily; improved versions were then tried but the zip did not catch on until the 1914–18 War when the American armed services starting using zips everywhere since they were more convenient for the soldiers than buttons.[40] The world's first cannery was established in 1812 in Bermondsey, London; in those days the contents were introduced through a hole at the top of the can and sealed with a soldered disc. But the direction on the label said: 'Cut around the top near the outer edge with a chisel and hammer.' This was a considerable impediment and the tin opener was soon invented.[40] In 1847, Fry and Sons in Britain made the world's first solid eating chocolate. Daniel Peter, a Swiss, did not find it sweet enough and, just by intuition, added milk to it. This was the original milk chocolate—and the beginning of Swiss chocolate-making supremacy.[41]

But perhaps the oddest story is this: towards the end of the nineteenth century there were two undertakers in Kansas City. One of them was a Mr Almon Strowger. His rival's wife was an operator on the local telephone exchange who consistently diverted Mr Strowger's calls, and hence his business, to her husband. To overcome this handicap, Strowger developed the basic design of the mechanical telephone exchange (still named after him) and these clicking/clacking exchanges have been in operation for many decades and will remain in use until the last Strowger exchanges have been replaced by automatic ones.[42]

'Law and order' can also play a part in technological advance. The elimination of piracy in earlier centuries made it possible to dispense with the carrying of heavy metal deck-guns on merchant ships; this change not only vastly increased the freight-carrying capacity of ships but improved the safety of their design by lowering the centre of gravity.[43]

Notes

1. Apart from the classical views of Malthus and Ricardo, some more recent and also more specific works may be mentioned, such as—among others—that of W. S. Jevons, *The Coal Question*, London, Macmillan, 1865; *Resources for Freedom, A Report to the President by the Materials Policy (Paley) Commission*, Washington, US Government Printing Office, 1952; J. W. Forrester, *World Dynamics*, Cambridge, Mass., Wright-Allen, 1971; or D. L. Meadows *et al.*, *The Limits to Growth*, London, Earth Island, 1972. For a thorough survey and critique of most of these works, see K. L. R. Pavitt, Malthus and other economists in H. S. D. Cole, C. Freeman, M. Jahoda and K. L. R. Pavitt (eds), *Thinking About the Future*, London, Chatto & Windus, 1973, pp. 137–58; for a shorter survey, G. F. Ray, 'Raw Materials, Shortages and Producer Power', *Long Range Planning*, **8**, 4 (August 1975), pp. 2–17.
2. C. Freeman, 'The Plastics Industry: A Comparative Study of Research and Innovation', *National Institute Economic Review*, 26, November 1963, pp. 22–62; 'Chemical Process Plant: Innovation and the World Market', *National Institute Economic Review*, 45, August 1968, pp. 29–57; and Chap 3 on 'Synthetic Materials' in *The Economics of Industrial Innovation*, Harmondsworth, Penguin Books, 1974, pp. 74–107.
3. George F. Ray, 'Mineral Reserves: Projected Lifetimes and Security of Supply', *Resources Policy*, **10**, No. 2 (June 1984), pp. 75–80.
4. OECD, *Interfutures—Facing the Future*, Paris, OECD, 1974, p. 41.
5. 'A Programme for the Wider Application of New and Improved Materials and Processes', Report of the Collyear Committee, London, HMSO, 1985; engineering ceramics, pp. 19–20; composite materials, pp. 13–18; thermoplastics, p. 14; metal matrix p. 17.
6. Among others, N. Rosenberg, *Perspectives on Technology*, London, Cambridge University Press, 1976, particularly Chapter 14, pp. 349–59; and *Inside the Black Box: Technology and Economics*, London, Cambridge University Press, 1982, pp. 55–80.
7. Freeman writes on the 'professionalisation' of R & D in *The Economics of Industrial Innovation*, (op. cit., note 2, p. 48) and finds that one of the reasons

for the success of the German chemical industry was that the leading companies 'were among the first firms in the world to organise their own professional R & D laboratories'. Elsewhere ('The Plastics Industry', op. cit., note 2, p. 32) he adds, however, that 'gifted individuals still play an extremely important part in the inventive process'.

8. For example, the OECD report, 'Interfutures', (op. cit., p. 45) states 'For asbestos, there are as yet no substitutes in many applications . . . technological advances before the end of the century in the production of synthetic asbestos substitutes would therefore be highly desirable, particularly since asbestos poses environmental problems'. See also G. F. Ray, 'Research Policy and Industrial Materials', *Research Policy*, **8** (1979), pp. 80–92.
9. Just one example: Freeman ('The Plastics Industry', op. cit., note 2, pp. 22–3) showed that the locus of production and exports of new plastics in advanced countries was more a function of technical progress than of relative factor costs.
10. W. H. McNeill, 'Coping with an uncertain future—A Historical Perspective' in C. J. Hitch (ed.), *Resources for an Uncertain Future*, Baltimore, Johns Hopkins, 1978, pp. 59. (This was a collection of papers presented at a forum marking the 25th anniversary of Resources for the Future, Washington, established in 1952.)
11. Studies specifically discussing 'new' materials or materials innovations in general are relatively rare (in contrast to those concerning particular materials). The four-volume *Science in History* by J. D. Bernal (Cambridge, Mass., MIT Press, 1971) discusses many of the materials now in use from a scientific angle, while D. S. Landes analyses the development of many of them from the standpoint of an economic historian in *The Unbound Prometheus—Technological Change and Industrial Development in Western Europe from 1750 to the Present*, London, Cambridge University Press, 1969. The case histories of selected new materials are contained in J. Jewkes, D. Sawers and R. Stillermann, *The Sources of Invention*, London, Macmillan, 2nd edn., 1969. A wide-ranging survey, shorter but more to the point, is in G. F. Ray, 'The contribution of science and technology to the supply of industrial materials', *National Institute Economic Review*, 92 (May 1980), pp. 33–51 and, covering a limited field, in C. Freeman, 'Synthetic Materials', op. cit., note 2, pp. 74–107.
12. G. F. Ray, 'The Wider Horizons of 'Industrial' Innovation' in M. J. Baker (ed.), *Industrial Innovation—Technology, Policy, Diffusion*, London, Macmillan, 1979, pp. 1–14.
13. Ray, op. cit., note 11. Stripmining, p. 38; nitrogen (Haber), 41; glycerine, p. 42; rubber, p. 42; acetone, p. 42; paper, p. 43; sulphur, p. 44; camphor, p. 44; man-made fibres, pp. 45–6; rare earth metals, p. 47; glass fibre, p. 47; aspirin, p. 46.
14. Sir Kingsley Dunham, 'How Long will our Minerals Last?', *New Scientist*, 17 January 1974, pp. 1–3.
15. According to Nordhaus, the price of copper relative to the hourly wage rates in American manufacturing declined from around 800 (index, 1970 = 100) to 82 in 1960. S. D. Nordhaus, 'Resources as a Constraint on Growth', *American Economic Review*, **64**, (1974), pp. 22–6.
16. Landes, op. cit., note 11; Thomas process, p. 352; sugar-beet, p. 144; alkalis, pp. 110–11; paper, p. 269; sulphur, p. 203; bakelite, p. 276; dyes, pp. 274–66; fertilizers, p. 515.

17. B. Gold, W. S. Peiree, G. Rosegger, M. Perlman, *Technological Progress and Industrial Leadership*, Lexington, Heath, 1984, pp. 309–11.
18. J. Platt, *British Coal*, London, Lyon, Grant and Green, 1968, pp. 8–11. G. F. Ray, 'Energy Economics: a Random Walk in History', *Energy Economics*, **3** (1979), pp. 139–43.
19. C. R. Fay, *English Economic History*, Cambridge, Heffer, 1948, pp. 111–12; J. L. and B. Hammond, *The Rise of Modern Industry*, London, Methuen, 1925, pp. 179–89.
20. D. H. Wallace, *Market Control in the Aluminium Industry*, Harvard Economic Studies, Vol. LVIII, Cambridge, Mass., Harvard University Press, 1937.
21. K. C. Li and C. Y. Wang, *Tungsten*, American Chemical Society Monograph No. 94, New York, Reinhold, 1955.
22. Rosenberg, *Perspectives*, op. cit., note 6, pp. 213–79, gives many illustrations for materials and also for 'new' processes, such as the oxygen method of steel-making.
23. B. G. Reuben and M. L. Burstall, *The Chemical Economy*, London, Longman, 1975, pp. 17–18.
24. Jewkes *et al.*., op. cit., note 11; fats 256–7; detergents 304–7; tungsten carbide 319–21; titanium 314–17; penicillin 278–9.
25. F. Höllscher, *Kautschuke, Kunststoffe, Fasern*, Ludwigshafen, BASF, 1972, pp. 23–34.
26. There were important political consequences of Weizmann's development work that contributed significantly to the wartime security of the supply of acetone. Weizmann turned down offers of honour or financial reward and was eventually recognized (as one of the leaders of the Zionist movement) by the Balfour Declaration in 1917, leading to the establishment of the state of Israel, of which Weizmann became the first president. Thus, Weizmann's contribution was an important one among the many other, chiefly political, considerations that eventually resulted in the Balfour Declaration.
27. *Encyclopaedia Britannica*, 1974 edn. 'Weizmann', Vol. X, p. 605; 'paper/wood pulp' pp. 13, 966; 'rare earths' VIII, p. 422; 'glass fibre', IV, p. 568.
28. Rosenberg gives many examples for innovative responses to materials shortages in *Perspectives*, op. cit., note 6, pp. 249–59.
29. Reuben and Burstall, op. cit., note 23; bakelite, p. 29; aspirin, pp. 17, 350.
30. Freeman, op. cit., 1974, note 2; dyes, pp. 48–51; plastics, pp. 74–83; cellophane, p. 94.
31. J. E. Tilton, *The Future of Non-Fuel Minerals*, Washington, Brookings, 1977, p. 22.
32. 'The Incredible Crystal: Diamonds', *National Geographic*, **195** (1979).
33. US Bureau of Mines, *Mineral Commodity Summaries, 1983*, Washington, US Government Printing Office, 1983; diamonds, pp. 44–5; quartz, pp. 122–3; rare earth metals, pp. 124–5.
34. C. Freeman, 'The Determinants of Innovation', *Futures*, **11**, No. 3 (June 1979), p. 206.
35. Rosenberg, *Perspectives*, op. cit., note 6, p. 197.
36. James M. Utterback, 'The Dynamics of Product and Process Innovation in Industry', in C. T. Hill and J. M. Utterback, *Technological Innovation for a Dynamic Economy*, New York, Pergamon, 1979, pp. 46–7.
37. The importance of the second 'D' in the R & D &D process is often overlooked. See C. Freeman, *Design and British Economic Performance*, London, Royal College of Art, 1983.

38. 'The Fellowship of Engineering', *Modern Materials in Manufacturing Industry*, London, 1983; carbon fibres, pp. 40–6; ceramics, pp. 49–51, 62–63; Sialon, pp. 47–8.
39. C. Freeman, C. Cooper and K. Pavitt, 'Policies for Technical Change', in C. Freeman and M. Jahoda, *World Futures: The Great Debate*, London, Martin Robertson, 1978, p. 209.
40. *Reader's Digest*, April 1982, p. 75.
41. *Reader's Digest*, April 1982, p. 16.
42. *Financial Times*, 29 April 1985, p. 16.
43. H. G. Johnson, *Technology and Economic Interdependence*, London, Macmillan, 1975, p. 10.

3 The State and private enterprise in high technology industries

Richard Nelson

Governments in the countries of Western Europe, the United States and Japan are grappling with the question of the kinds of programs they should mount to enhance the competitiveness of their high-technology industries. The discussion in Japan may be less intense, because the Japanese appear to have worked out some relatively well-defined policies in this area, while the other advanced nations are still groping. What the Japanese model actually is, and whether or not it is importable, is an important part of the policy debate in Europe and in the United States. Policy towards high-technology industries is clearly one of the hot topics of the 1980s.[1]

This essay, however, will not be particularly topical. Rather, my focus is on past experience with government programs that have supported R & D in industry. My objective is to illuminate the different rationales that have been used to support such programs, and the different kinds of programs, and to begin to assess what kinds of programs seem to work well, or poorly, in different contexts.

The question of the appropriate governmental roles to support of industrial R & D, while now a fashionable one, has been very little researched. Economists, and other scholars, studying the processes of technological advance in industry, have tended to stress the role of private firms, proprietary knowledge, and the stimulus provided by a competitive market environment. The influence of Schumpeter is apparent in much of the recent research.[2] However, Schumpeter was surprisingly blind to the considerable public roles in inducing and supporting technological advance in industry. And a few modern scholars have been inclined to look closely at that part of the picture.

There have been, of course, a number of relatively detailed studies of particular government policies in support of industrial innovation.[3] In my own analysis, I draw heavily on these. However, with few exceptions, these studies have concerned themselves little or not at all with what private firms were doing in the industry in question, on their own initiative. A general theme I shall develop is that government policies in support of R & D in industry have to be understood in terms of how these interact with private policies. Also, the case studies in question tend to be of particular policies. There has been little in the way of effort to try to pull together the case study evidence, to identify what is similar and different about the particular cases, and to sketch out a general picture of the landscape. This is what I will try to do here.

My own research is based both on the secondary sources mentioned above, and on some primary research. It reveals a rich and variegated picture. Governments have tried many different kinds of R & D support programs. Some of these appear to have been very successful, given the objectives. Others have been frustrating failures, or worse. The heart of this paper will consist of my description and analysis of some of this diversity. However, it is useful to begin by discussing some general features of technological advance in competitive industries.

Characteristics of technological advance in capitalist economies

In this section I highlight three characteristics of technological advance in market economies.[4] These seem to hold, if in different degrees and forms, in every industry and in all the major countries. They are, first, that technological advance is a business involving major uncertainties. Second, a competitive market context, which is an important part of the system generating technologial advance in most industries, has certain strengths and weaknesses, and imposes both opportunities and constraints on governmental roles. Third, there are several different kinds of public R & D support programs; each is appropriate to further certain purposes, but not others, and each is limited in the range of industry structures where it is applicable. All of this bears on the question—what kind of policies are likely to be feasible and effective in spurring technological advance in industry—although the opportunities and constraints differ from industry to industry.

Uncertainty

It is important to recognize the essential uncertainties which surround the question—where should R & D resources be allocated—in an industry where technology is advancing rapidly. There are generally a wide number of ways in which the existing technology can be improved, and at least several different paths toward achieving any of these improvements. *Ex ante* it is uncertain which of the objectives is most worthwhile pursuing, and which of the approaches will prove most successful. Before the fact, aviation experts disagreed on the relative promise of the turboprop and turbojet engines; those who believed in the long-run promise of commercial aircraft designed around turbojet engines were of different minds about when to go forward with a commercial vehicle. Whether and when computers should be transistorized was a topic on which computer designers disagreed; later, the extent and timing of adoption of integrated circuit technology in computers was a subject which divided the industry.

There are market as well as technological uncertainties. It is no easy task to judge how much merit customers will see in a radically new design. The customers may not know themselves before they have tried it out. The favorable public response to the smoothness of jet passenger flight was easy to underestimate, and the lack of willingness to pay for supersonic flight easy to have miscalculated. Before such machines were made available, there was no apparent business demand for computers. The value of an innovation may depend upon unpredictable events, such as whether a complementary product is available, or how the market develops for a product for which it is a component part. The post-1973 hikes in fuel costs surely hurt Concorde, and helped Airbus.

If the problem were simply uncertainty, but everybody agreed on the structure of the uncertainty, one could define the R & D allocation problem as being something like a dynamic programing problem involving uncertainty and learning. An optimum strategy in such a context may well involve exploring a variety of different possibilities, and holding off commitment to a single one until lots of evidence is acquired. There is considerable merit to this perception.

But a key characteristic of the R & D environment is differences of opinion and vision. Human beings, and organizations, seem to be innately limited in the range of things they can hold in mind at any time, and even in the way they look at problems. Some individuals simply see things about a problem, or about a possible solution, that others do not see; what is seen may or may not actually be there. The fact that different people look at a problem in different ways and see different things about it means that terms like insight, creativity, genius, are often applied to successful inventors or laboratories. And, at any time, there is inevitable disagreement about what is the best course to take, and about who will be right about the matter this time, even if one could get open, unguarded discussion of the question. Committees of experts are unreliable judges of these issues even if, or particularly if, they are forced to arrive at agreement.

The implications are important. The uncertainty that characterizes technological advance in high technology industries warns against premature unhedged commitments to particular expensive projects, at least when it is possible to keep options open. The divergences of opinion suggest that a degree of pluralism, of competition among those who place their bets on different ideas, is an important, if wasteful, aspect of technological advance.

The competitive market context

As noted, Joseph Schumpeter, more than anyone else, has shaped the way scholars view competition in technologically progressive industries.

Schumpeter's core message was that the most socially valuable form of competition, in capitalist economies, was through technological innovation.

It is proprietary technological knowledge that drives the capitalist engine. The principal ways to achieve proprietary benefit are secrecy, patent protection, and through a head start. There are significant differences among technologically progressive industries in the extent to which these different mechanisms are effective. But while the mechanisms differ, what is noteworthy is that, in every industry where privately funded R & D has been substantial, firms are able to profit from their R & D successes through one mechanism or another.

The Schumpeterian system has been an extraordinarily effective engine of progress. It has shown sensitivity to changing patterns of demand by consumers. The pay-off to a firm lies not simply in producing a technologically advanced product, but a product that consumers will buy in quantities at a price that is profitable. Profitable companies and technologically progressive industries are characterized by strong market research, as well as by strong R & D. At the same time, competition among firms, accompanied by secrecy about just where each is laying its technological bets, willy nilly generates a reasonable diversity of approaches to problems and new products offered to the market for selection.

However, a careful scrutiny either of the models that capture, in abstract form, the nature of Schumpeterian competition, or of the empirical history of technological advance in any field, indicates that the portfolio generated by market competition can in no way be a clustering of effort, verging on duplication, on alternatives widely regarded as promising, and often a neglect of long shots that, from society's point of view, ought to be explored as a hedge. The premium placed on achieving an invention first, so as to get a patent, or at least a head start, may lead to undue haste and waste and duplication of effort. That three companies—McDonnell-Douglas, Lockheed, and the Airbus consortium—all tried to compete in the market for wide-bodied, medium-sized airliners surely meant that total costs were excessive, if it also meant that the airlines got a good deal.

On the other hand, the fact that certain kinds of technological advances are not well protected by patents, and are readily copies, deters companies from investing in these, even though a significant advance would lead to enhanced efficiency or performance. Before the advent of hybrid corn seeds, which cannot be reproduced by farmers, seed companies had little incentive to do R & D on new seeds, since the farmers, after buying a batch, could simply reproduce them themselves. The farmers themselves had little incentive to do such work since each was small and had limited opportunities to gain by having a better crop than a neighbor. Within an industry, different kinds of problems vary in the extent to which the problem solver gains a special advantage. In an industry where scientists and engineers are mobile it is hard to keep secret for very long information about the broad operating characteristics of a particular generic design, or about the properties of certain materials. Such knowledge is generally not patentable and, if patentable, would be very hard to police.

It is tempting to regard these kinds of 'market failures' as providing both justification and guidance for governmental actions to complement, substitute for, or guide private initiatives. At the least, their recognition guards against the simplistic position that the R & D allocation naturally induced by market forces is in any sense 'optimal'. However, propositions about where and how market forces work poorly cannot alone carry the policy discussion very far.

Market institutions themselves constrain public policies. The fact that much of technological knowledge is proprietary is an important constraint. In the general run of things, a company will not willingly disclose to its competitors, or to a public agency, the way it thinks the technological bets ought to be laid. As a result, a government agency may be cut off from the most knowledgeable expertise on the question. While, as will be elaborated later, a portion of relevant technological knowledge is public, the details of what works well, and what the key problems are, may be known only to the firms in the industry, and perhaps their customers. Market knowledge may be very difficult for a government agency to obtain, unless the companies want to give it. Relatedly, a government agency may be sorely limited in its ability to find out where private companies are allocating their own R & D efforts. To the extent that public monies aim to 'fill gaps' in the private R & D portfolio, it may not be easy to identify where these gaps are. There is also a danger that public funds may duplicate, or replace, private funds.

Also, private firms are likely to resist governmental programs that they see as cutting into their own turf, or as likely to aid competitors. It is generally a mistake to think that an industrial policy can successfully be imposed upon an industry. To be effective, a policy requires a degree of cooperation and participation from the industry, and members of the industry are inevitably going to be influential in shaping any policy. This is a different thing from saying that the firms in the industry will inevitably control a policy. However, they need to be understood as active players in the game, and not as passive acceptors of government policies and orders.

Bases for government R & D support

I have found it useful to think in terms of three roughly distinguishable kinds of R & D programs directed towards enhancing the technological competence of industry: those that mainly support generic research, those that are basically concerned with government procurement, and those that explicitly aim to promote commercial capabilities. These different kinds of programs accommodate, or avoid, or get through, informational and political constraints on government industrial R & D support in different ways.

Support for scientific and engineering research and education, largely at universities, has long been a recognized government responsibility. Much of this publicly-supported activity goes on in the standard academic basic scientific departments, like physics, which do not have tight links to particular technologies or industries. But a good portion of governmentally-supported university-based research and teaching is in the applied sciences—like pharmacology, or computer science, or electrical engineering—which tend to be quite close to certain technologies and industries.

The existence of these applied scientific disciplines partly reflects, and partly assures, that technological knowledge has an important public component as well as a private one. The public part of technological knowledge generally does not relate to the design or operational details of a particular product or process, but to 'generic' knowledge—broad design concepts, general working characteristics of processes, properties of materials, testing techniques, etc. Such knowledge is often not patentable. While it can sometimes be protected by industrial secrecy, this may be difficult. Also, this is the kind of knowledge that must be imparted to those trained to be engineers, or advanced technicians. Therefore, it would seriously interfere with the ability of technical schools, and universities, to provide good training if the relevant knowledge were proprietary.

Research in the applied sciences is conducted by scientists and engineers in industry, as well as in universities. In some fields, as now seems to be the case in certain areas of semiconductor and computer technology, the industrial R & D groups may be doing more advanced work than the academics; in other cases industry-based research is conducted largely to provide a window into academic research. But, in any case, good communication between industrial scientists and academic scientists is an important part of the enterprise. The journals generally receive contributions from both. The scientific societies include both.

The presence of well-established networks of generic research scientists, drawn from industry as well as academia, provides one important base for government research support. So long as the R & D support program sticks close to generic work, the problem of proprietary rights is partially averted. A consultative structure already stands to help map out sensible allocations.

Public procurement demands are another traditional source of public involvement in industrial R & D. From far back in history, sovereigns have maintained arsenals and other warshops producing the goods they needed, and have concerned themselves with the adequacy of supply of military and other items. While defense is the largest procurement interest, in several countries space agencies, telecommunications networks, electric utilities, and airlines are government-operated and controlled, and are also important sources of demand for high-technology industries. A strong procurement interest at once yields political legitimacy to a program, and is usually associated with in-house governmental technological expertise. Also, a government agency has some ideas about what kinds of products *it* would buy.

While procurement-orientated policies may involve support for generic research to enhance general design capabilities in a particular field, such policies tend to be more narrowly focussed than more general generic research support policies, and to be aimed at developing and bringing into production particular products. In contrast to more broadly orientated generic research support programs, where there is generally considerable reliance on the relevant scientific or technical community for guidance, procurement-orientated policies tend to be tightly controlled by government agencies pursuing their own ends as they see them.

Policies that stress generic research, or procurement, do not directly engage the state in attempting to assure the viability in commercial competition of particular industries, firms, or products. But many government R & D support programs are explicitly aimed at promoting commercial goals. While support of generic research and technical education may be part of an industry promotion program, such programs tend to focus quite directly on enhancing the commercial competence of national firms. The commitment to a particular industry is often connected to a procurement—generally national security linked—interest, but in the promotional programs the aim is not particular procurement objectives but to develop or preserve a commercially viable industry. Indeed, in a number of such programs that will be considered here, procurement might be better regarded as an instrument used to further the industry, than vice versa.

Policies expressly aimed at promotion of commercial competence vary in the extent to which the industry itself provides support and guidance for the program, and government in effect facilitates, coordinates, and funds, and the extent to which a government agency itself lays out the agenda and the firms largely follow a government lead. Programs which are guided by industry both tap industry expertise and are supported politically by the industry. Such programs also tend to be constrained by the interests of the firms in the industry, as they see them. A program in which the government leads, on the other hand, requires that somehow the government make judgments about the nature of the market and competition in the industry, as well as about technological matters. But such a program does give the government capability to plan the evolution of technology, for better or worse.

Many government R & D programs of course involve a mix of purposes and activities. An interesting question is, when there is such a mix, do the components complement each other, or do they clash? In the remainder of this paper I consider the efficacy of different kinds of programs in terms of the extent to which they enhance commercial strength.

A sample of publicly-funded promotional R & D programs

In this section I describe and analyze the following collection of programs: American agriculture; French, Japanese and American support of semiconductors and computers; American, British, and French support for civil aviation; and the nuclear power reactor programs of these same countries. There are certain similarities among these programs, and also some important differences. The differences are associated with, first, the extent to which enhancing commercial capability was an explicit objective and R & D funds were targeted to that purpose; second, the strength of government procurement or other national security interests in the industry; and third, the nature of commercial competition among the firms in the industries.

United States agriculture[5]

Support by the federal and state governments for agricultural research goes back over 150 years. By the mid-nineteenth century, a number of individual states were supporting agricultural experimentation stations, and the federal government was operating one under the auspices of the patent office. The Land Grant College Act of 1864, which established funding for state universities orientated toward the agricultural and mechanical arts, and the Hatch Act of 1887, which provided for federal funding of agricultural research in each state, established the basic legal structure and institutional apparatus for the federal–state cooperation which still exists.

The current system consists of both federal (United States Department of Agriculture) and state units. Most state units are integrated with land grant universities. Many of the researchers hold university teaching positions. While there are still a few freestanding federal experimentation stations, federal programs have increasingly been melded with state ones.

In terms of the variables discussed earlier, the agricultural research support program involves a blend of generic and promotional elements. There is no link to any obvious governmental procurement interest, although, ironically, the federal government has been drawn into making markets for agricultural products produced by the enormously productive system its R & D support has helped create. The program, from the beginning, has been called forth by political pressure from farmers, and guided through mechanisms that we shall consider shortly.

Intellectually grounded in such traditional fundamental sciences as chemistry and biology, the system of publicly-supported agricultural research has given birth to a large number of applied sciences, such as plant breeding, and animal husbandry. Much of the work within these fields is generic, in that research is aimed at enhancing understanding relevant to a wide range of problems. These applied sciences have special journals, textbooks and other standard academic trappings. They form an intellectual bridge from the fundamental sciences to the practical problems of agriculture that can be attacked by science. They, in effect, provide the operating scientific basis for research aimed at finding out, for example, just the appropriate amount and timing of fertilizer applications for a particular kind of crop, or the causes, preventions, or cures, for a particular disease. Historians of technological advance in agriculture have argued that the applied research and development efforts of the experimentation stations did not yield particularly high rates of return until there was in place a relatively solid scientific understanding of the relevant underlying phenomena. Thus, modern hybridization did not go forward effectively until Schull discovered, and understood, that to get good hybrids one generally needed to start with pure varieties.

The typology sketched earlier suggests that public support for generic research relevant to agriculture should be looked at somewhat differently from public support for research of direct and immediate relevance to farmers. The agricultural generic research system has much in common with other generic research systems orientated toward particular industries or technologies. Governments have provided support for such generic research systems, across a wide front of industries. However, in such fields as medicine, the publicly-supported part of the system (in the United States financed largely through the National Institutes of Health) has rarely ventured beyond generic research. The applied research and development which has a significant chance of leading to proprietary products, has been left largely to private industry.

What is it about farming that has led the ‘firms’ in this industry to actively demand, not just resist, government R & D support for applied work, as well as generic? One factor is that individual farmers, unlike individual firms in many manufacturing industries (like pharmaceuticals), are not in rivalrous competition with each other. It does not hurt one farmer if his neighbor becomes more productive. Put another way, there is very little proprietary knowledge among farmers. At the same time, regional groups of farmers perceive that their sales can be enhanced and their profits increased if their productivity and the quality of their crops are improved. Thus, a regional group of farmers has a strong interest in getting effective applied research and development undertaken, aimed to enable them, as a group, to become more competitive.

At the present time, state financing of agricultural research is significantly larger than federal financing. The size and character of a state program is effectively moulded by the agricultural interests in the state. Experimentation station directors have to justify their budgets in state legislatures, within which farmers carry considerable influence. While the federal part of the system provides some mechanism for coordination, and for supporting work for which no partiular state has an especially sharp interest, the system as a whole is very decentralized. Some current critics of the system, speaking from the vantage point of the US Department of Agriculture, or from the basic scientific community, have complained about duplication of effort, and lack coordination. They have also observed that some of the research is of low quality, which often has the meaning of being too dominated by imperatives of particular practical objectives. However, the governing structure has certainly made the system responsive and accountable to farming interests as the farmers see them.

Of course, public experimentation stations are not the only source of new or improved agricultural technology. There has long been a tradition of private invention and innovation of agricultural machinery. In recent years a considerable amount of R & D has been done by private producers of seeds, pesticides and fertilizers. By and large, the public part of the system has not encroached much upon the private part of the system, but rather has worked in areas where private capabilities or incentives are weak. Again, the political mechanisms of control seem to be the reason. On the one hand, companies which are doing R & D in a particular area can and do complain and lobby the state legislatures when their turf is being encroached on. On the other hand, where farmers deem the efforts of private companies to be satisfactory, there is no particular pressure for public programs in that area.

A large number of studies have shown that this public R & D support system has yielded a high social rate of return. The primary beneficiaries have been consumers, and probably not farmers. As productivity has increased, costs have fallen. Farming has been a sufficiently easy industry to enter so that prices have fallen with costs. Indeed, since agriculture, as an entire industry, faces a relatively inelastic demand curve for its products, rapid productivity growth accompanied by falling prices has forced a dramatic fall over the years in the number of people who can earn an average standard of living in it. At the same time, the enormous productivity of American agriculture has enabled the United States to continue to be a major exporter of agricultural products, despite our very high wage rate relative to other countries. While plentiful rich land is part of the explanation, our agricultural research and development support program is also a very important part of that story.

American, French and Japanese electronics[6]

My next set of case studies is one of semiconductors and computers. The contrast with agriculture is striking. On the one hand, commercial competition in semiconductors and computers is fiercely rivalrous, proprietary information is very important, and companies keep their own bets a closely held secret. On the other hand, these industries are central to defense procurement, and, more generally, governments have viewed having a strong domestic industry as essential to economic as well as military independence. The former factor makes problematic the kind of publicly-supported, industry guided applied R & D arrangements found in agriculture. The latter means that the natural route for government involvement is through a procurement-like policy.

The Americans, French and Japanese have all had programs of R & D support for these industries, but these have differed in important respects. The American programs have been virtually exclusively procurement-orientated. They have not been expressly concerned with 'promotion' and thus will not be considered in detail here. However, they have made a very substantial contribution to the commercial competence of the American industries, and it seems likely that had this not been the case, a more expressly promotional policy would have been adopted to complement the procurement orientated policies. The French R & D support programs have explicitly involved both procurement and promotional objectives. Thus they provide one case for scrutiny regarding the question of what happens when procurement and promotional objectives are mixed, whether they mutually reinforce each other, or whether they tangle. Japan, in contrast, has developed R & D support programs that basically aim to promote commercial capabilities, although behind the scene the Japanese government has been concerned that Japan should not be dependent upon other nations for certain kinds of products. Thus, the Japanese program of promotional R & D support in these industries is, like the American agricultural case, a relatively pure one. As we shall see, however, the differences between the two industries in the nature of competition has had a profound effect on the nature of the feasible promotional programs.

The American semiconductor and computer industries, still arguably the strongest in the world, were enormously helped in their early days by a Department of Defense interest in the underlying technologies.[7] While the details differ, the broad stories in the two industries are similar.

Virtually all of the exploratory research and development efforts that led up to the early electronic computers was financed by the armed forces. The government was almost the sole market for the early operational computers and continued to be the dominant market into the early 1960s. Governmental funding of R & D and procurement was motivated strictly by national security interests. There is not a hint that anybody in government had in mind that they were creating an industry that would be a major economic asset. Very few of the companies involved in the early work for government believed that there would be a large civilian market, as well as a government one. Of course, it later turned out that there was a very large non-governmental market for computers. The massive government support to computer technology provided American companies with a head start that has still not been overcome by foreign companies.

The American experience with semiconductors has some similar elements and some differences. Perhaps the key difference is that the bulk of the early R & D was privately, not publicly, financed. The work leading up to the transistor was motivated by perceptions of the value of such a device for the telephone system. Once the transistor had been invented, however, the Department of Defense very quickly understood the potentials of the new technology for military hardware. As with the case of computers, government support was motivated by a procurement interest, not an interest in establishing an industry that would be a national economic asset. Yet, as with the case of computers, the latter was one of the results.

After the initial breakthroughs, the American military programs, complemented by the NASA efforts in the decade overlapping the 1960s and 1970s, had two striking characteristics. First, considerable generic research was supported. Most of this generic work was done in industry and enhanced the broad design capabilties of the companies involved. Second, the particular items that DOD and NASA contracted to procure often required state-of-the-art advance, and the efforts pulled computer and semiconductor technologies to capabilities never before achieved. Many of these capabilities had civilian applications. Several observers have proposed that, by the mid-1970s, American military procurement demands were no longer calling forth advances in computer and semiconductor technology that led to significant commercial advantages for the engaged American firms. Rather, the civilian computer industry was by then the principal instigator of technological advances. If true, the era of American dominance in these industries may be over, or at least the lead may be challengeable by countries without as massive or ambitious a military program.

The Japanese experience suggests this may be so.[8] The rapid takeover of the American color television market by Japanese manufacturers in the late 1960s came as a shock to many Americans, and was rightly regarded as an indicator that American preeminence in consumer electronics was under threat. While there were many factors behind the Japanese success in this field, at least part of the reason was a MITI funded cooperative research program that enabled Japanese companies to get ahead of American companies in exploiting the opportunities afforded by integrated circuits. The support here was for generic research, not for the design and development of specific products. The Japanese companies themselves initiated and funded product design and development.

Japanese policies in direct support of their semiconductor industry have a similar flavor. During the early 1970s, MITI organized and partially funded various semiconductor R & D projects. In the mid-1970s the VLSI (very large scale integration) research program was initiated. This program, as with the earlier one directed toward color television technology, was largely generic in nature. While a large number of patents came from that program, the basic purpose and result of the program was to bring Japanese companies up to the state of the art along a rather wide front. It took the form of several cooperative research projects that were staffed, to a large degree, by company scientists and engineers. At the same time, Nippon Telephone and Telegraph sponsored considerable R & D associated with its procurement efforts.

The case of computers is somewhat special because of the presence in Japan of IBM. IBM got into Japan before the war and its leverage on the Japanese was also enhanced by the fact that it held some of the basic computer patents. MITI successfully bargained licenses out of IBM, and got IBM to agree to limit its Japanese sales but IBM remained, until 1981, the largest computer company in Japan, at which time she was surpassed by Fujitsu.

It would appear that in the late 1960s MITI made a judgment that Japanese computer capability was too fragmented, and that merging would be in order. The large Japanese electronics companies proved unwilling to separate out their computer design and manufacturing capabilities and to merge these. As a compromise, MITI organized and helped support several different research and development groups, each group orientated around a particular strategy for computer design and commercialization. The target for these efforts was not a government market which could be assured and shared by the cooperating firms, but the highly competitive general commercial market. Because of this, the cooperative R & D arrangements often proved fractious since the work being done touched on the potential proprietary interests of rivalrous firms.

More recent Japanese programs have stressed more generic research. Unlike the earlier program, the fourth and fifth generation computer programs do not appear to involve particular companies in commitments regarding the nature of the computers they will ultimately design and market. What MITI now appears to be trying to do is not to direct the commercial developments of computers in Japan, but to see that the Japanese companies have the technological capabilities to compete with IBM and the other major Western companies in designing and developing the next generation of computers.

The three cases considered up till now—American agriculture, American electronics, and Japanese electronics—are all relatively 'pure' ones, in the sense of the categories developed earlier. I turn now to a very 'mixed' case—that of French government R & D support for its computer and semiconductor industries.

French R & D support programs in these two industries involved a close mix of procurement and promotional interests.[9] I shall not discuss here the French military R & D programs in any detail, save to flag an important fact. French military R & D spending has been very limited compared with the American programs. Relatedly, it seems clear that the French military programs have neither been so ambitious about making state-of-the-art advances in particular procurement efforts, nor have they supported as wide a range of generic research aimed to open up new technologies, as the American programs (or recently, the Japanese). As a result, French weapons technology has to a great extent tracked American technology. The French have developed some very good equipment, but the companies that did the R & D did not in general learn to do things the Americans could not do. The key component technologies—in this case, computers and semiconductors—have not been pulled along so far by military procurement and R & D in France as they have in the United States. As a result, the French have felt themselves lagging in these areas, and have been concerned with the effect of this on their ability to produce first-class military equipment.

Indeed, modern French programs in support of these industries have recognizable origins in French frustration at the refusal of the American government to let the French buy from an American company a large computer needed for its nuclear programs. This occurred twenty years ago. As a result of this experience, and subsequent ones, the French government committed itself to the establishment of a French computer design and production capability that would, among other things, enable France to meet her military needs. By this time the commercial market, with both private and governmental customers, was large and growing. It was natural to think of government support for the French industry as furthering commercial capabilities, as well as the particular ones judged essential by the French military and other government procurement interests. A similar dependency upon American companies for integrated circuits for military equipment, and for computers, led the French government to a similar commitment regarding a French semiconductor industry.

Both French tradition, and the centrality of national security-related procurement interests, made it natural to proceed by identifying, or establishing, particular firms as 'national champions' with the responsibility both to meet procurement needs and to develop a commercial capability. French military demands were judged not to be large enough to warrant the maintenance of several competing firms, and in any case the French faith in competition has always been much less strong than the American. Note, however, that by bundling its promotion program with its procurement program, the French government implicitly picked one company to carry its commercial banners, as well as to develop and preserve the relevant military procurement capabilities. The French in effect blocked themselves from the road followed by the Japanese of providing support and encouragement to a number of companies.

At the same time, this strategy broke down what would otherwise have been barriers to involvement by French government officials in the commercially orientated R & D strategies of its chosen French firms. Zysman has argued that the mixing of military and commercial purposes led to a schizophrenia that virtually guaranteed failure to achieve the latter objective. As noted, French military R & D spending alone was not large enough, nor were the objectives ambitious enough, to pull the technologies beyond where the Americans already were. At the same time, the military objectives, and the rhetoric associated with 'a national champion', led French government officials, who controlled various flows of funds to the companies, to encourage those companies to try to match the Americans where the latter were strongest. Various writers have remarked that government-pushed projects, while often allegedly for commercial purposes, have seldom been directed toward plausible markets, and that company-proposed projects have been judged on the basis of how they fit national, not necessarily commercial, objectives. Thus, the French companies could not hunt for commercial niches which could be developed into areas of major commercial strength.

There clearly are a wide variety of ways in which Japan differs from France, and the Japanese programs have differed from the French. Among other things, the Japanese were able to establish a closed domestic market for the products of their firms, and that domestic market was large enough to support a number of competing companies. Both because of its position within the EEC, the French civilian market could not be reserved for French firms. But by my reading of the evidence at least, this was not the major difference. The difference was that the French programs were simply not well designed to promote commercial competences, while the Japanese were.

American, British and French support for civil aviation

The story of government policies in support of civil aviation contains a number of elements in common with the electronics story. However, to a far greater extent than in electronics, governments—particularly the British and French—have financed the development and subsidized the production of particular designs aimed explicitly at a civilian market.

Except for the case of the supersonic transport, the United States government has been unique among the three in not involving itself in deliberate direct subsidization of civil aircraft development.[10] During the inter-war period, the government did take a direct interest in the development of the American aircraft industry. While motivated by the objective of having an industry capable of design and production of first-rate military aircraft, the policies aimed quite broadly to support the industry. The National Advisory Committee on Aeronautics was established in 1915 to 'investigate the scientific problems involved in flight and give advice to the military air services and other aviation services of the government.' As the statement of mission attests, the program was justified in terms of direct government (largely military) needs but, from the beginning, the problems NACA worked on were common to commercial as well as military aircraft. NACA's work on engine and airframe streamlining played an important role in enabling the design of the DC-3. That aircraft, and planes that evolved from it, dominated the commercial airliner market from the mid-1930s until the advent of passenger jet aircraft. During the 1930s, and after, the American government subsidized the airlines, and indirectly therefore, civil aircraft design and development, through contracts to carry airmail.

By the late 1930s, NACA began to concentrate more specifically on problems of special interest to the military, and the flow of civilian benefits diminished. After World War II, much of the 'generic research' mission which had been shouldered by NACA was shifted to the aircraft companies through DOD contracts explicitly with them. By the late 1950s NACA had been transformed into NASA and the orientation shifted largely towards space.

Even during the 1930s, the technological problems and opportunities of greatest interest to the military only partially matched priorities for commercial aviation development. However, until 1970 or so there was considerable overlap. During the post-World War II era, design and procurement of a new aircraft, or a new engine, for military use often led the advance of technology, with civil technology following. The American post-war preeminence in the commercial aircraft business arose directly out of military research and development and procurement contracts. The Boeing 707 was designed in parallel with a plane bought by the Air Force, and had many design elements in common. The American wide-bodied jets show their origins in military cargo planes and the engines that powered them. Until the supersonic transport episode, which I shall discuss later, there were no American government programs aimed expressly at helping in the development of commercial airliners, nor was there any pressure for such from the major aircraft producers.

The situation in Britain and France has been quite different.[11] In Britain, during World War II, a relatively explicit government plan was drawn up for post-war support of the design and development of civil aircraft. In the early post-war years several designs were in fact developed, according to the plan. Most of these efforts were absorbed short of a vehicle ready for a market test. The few that were fully developed turned out to be dominated by American aircraft.

It is interesting that the British-designed and built plane that marked the largest technological step forward—the De Havilland Comet—was developed and produced without government support. Turbojet aircraft were not in the plan. Comet, which got into production and use six years before the Boeing 707 and before the French built Caravelle, turned out to have fatal technical problems. Government funds did go into efforts at redesign, but these were not sufficient for the necessary modifications to be effected in time to beat out Boeing.

The experience of the British government in betting right was no better during the 1950s and 1960s than it had been in the immediate post-war period. During this time the government subsidized the design of more than a dozen aircraft. Only one—Viscount—can be regarded as close to a commercial success. The nationalized British airlines, BEA and BOAC, were in effect commanded to buy British planes and, as a result, were often disadvantaged relative to other airlines flying competitive routes that had freedom to shop around.

In the middle 1960s, partly in response to the financial losses being accrued, a committee was formed, headed by Lord Plowden, to consider the future place and organization of the British aviation industry. One of the committee's most important recommendations was that future efforts should be focused on collaborative efforts with other European countries. It was already clear that one such ongoing effort—the Anglo-French Concorde—was likely to be a financial albatross. However, the logic of the Plowden recommendation seems to have persuaded the British government that attempts to develop a purely national industry through subsidization and a guaranteed home market were extremely expensive and ultimately futile, and foreshadowed several cooperative ventures during the 1970s, notably Airbus.

The French story has some similar and some different aspects. By and large the effort has been more successful. During the 1950s the French government authorized the development of the turbojet Caravelle. The plane was designed for the short- and medium-range trips and, thus, found a niche in the first-generation jet market, where the other planes—707, DC-8, Comet—were designed for longer range. The Caravelle was dominated, however, by the Boeing 727 which appeared in the early 1960s. Except for the Caravelle, during the 1950s the French government did not really push or try to direct commercial aircraft design and development. Their efforts were focussed on military aircraft. There appears to have been little of the sense of urgency to establish or preserve a commercial aircraft industry that marked the British case, perhaps because during the war Britain had built up a large labour-force in her aircraft industry and France, of course, had not.

After Caravelle, the next major venture in civil aviation was the supersonic aircraft, the Concorde, a joint venture with the British. Enough has been written about the Concorde so that only a sketch is required here. In contrast to Airbus, which will be discussed shortly, in the case of Concorde very little attention was paid to the nature and size of potential markets, or to how sensitive those markets would be to price. Nor was the experience in military R & D heeded—that the cost of ventures aiming for a radical advance in technology tends to be greatly underestimated. The original $450 million estimate for development costs proved low by a factor of ten. Only the captive French and British airlines could be forced to accept delivery of the Concorde when it was finally ready for commercial operation in 1976, and both governments have had to subsidize the operation of the plane. Production was terminated in 1979. Only sixteen aircraft were produced.

The United States government was also drawn, or jumped, into subsidy and direction of a supersonic transport project. The American effort, which was begun several years after the European effort was launched, was a direct response to it, as well as a desire to exploit expected 'spillover' from the development of the B-70 strategic bomber prototype. Instead of the normal procedure in the development of specifications for a new commercial aircraft, in which there is significant interaction between the airlines and the company considering the venture, in this case the lead government agency—The Federal Aviation Administration—stipulated the performance requirements, with not much consultation with the airlines. Boeing won the contract competition. Serious technical problems (the original design proved unfeasible), cost escalation, and opposition from environmental groups, led to the program's demise in 1971. The experience with Concorde suggests the United States was lucky that the program never achieved a technically viable aircraft.

The Airbus case is an entirely different story, and, since not much has been written about it, warrants telling in some detail. As early as 1963,

Britain and France had begun discussions about a possible joint venture to produce a large commercial subsonic aircraft. By the mid-1960s, the Germans, who were eager to expand their presence in the aviation industry, joined the discussions. An agreement to start development on a 260 to 300-seat wide-bodied plane was signed in the fall of 1967. By that time the Douglas DC-10 and Lockheed L-1011 were aready under development. Both were planes of roughly this size, but aimed for mid-distance flight. The Airbus consortium tried to avoid competition with the Americans by choosing a two-engine design (the American planes each had three engines) tailored to the short-run market. This market niche was defined in discussions with the European airlines regarding the kinds of planes they would like to procure.

There were, and are, certain important features of the governance of Airbus Industrie. The top management of the firms involved has the authority to define both technical and marketing objectives for the project. While the participating governments hold the purse strings, and thus can ultimately veto decisions, government officials do not become directly involved in formulating design or marketing proposals. The top executives also have the authority over administration, and thus control how the decisions are implemented.

Despite a design apparently well aimed for a market niche (actually, by the late 1970s two designs), and despite a promising management system, during most of the 1970s the financial prospects for Airbus seemed dim. Through the late 1970s orders for Airbus were slim compared with those for the Lockheed and McDonnell–Douglas planes. Beginning in 1979, Airbus orders began to pick up dramatically. While it is still too early to tell if the consortium will make a profit, its planes have sold better than any other European-designed airliner ever made.

The fierce competition among the Airbus consortium, Lockheed with its L-1011, and McDonnell–Douglas with its DC-10, for roughly the same market reveals extremely sharply the conflicting nature of national policies in support of high technology industries for economic purposes. The American companies have complained, naturally, that foreign governments were heavily subsidizing their competitor.

Nuclear power[12]

In the field of nuclear power, the government of the United States, as well as that of the major European countries, has spent enormous sums of money over a long period of time with the objective of creating a commercially viable and internationally competitive power reactor industry. In all of these countries a special government agency has been explicitly charged with the job of guiding reactor development, and in several has done this in great detail. While by some standards the French and German programs might now be regarded as successful, these programs, as with the American and British, spent enormous sums of money on false leads. It is still not clear whether the overall rate of return on the programs will be positive.

However, the issues are complicated and tangled. In the first place, even more than in the case of aviation or electronics, policies in support of the development of nuclear power technologies have been tightly intertwined with explicit national security objectives. Second, in the early days of atomic power, concerns about environmental impact, and safety, were muted. As these concerns became better articulated, and represented in the political process, new design requirements and more stringent licensing procedures were imposed. The financial costs of nuclear power were thus significantly increased. Further, at roughly the same time that these factors were slowing the tide of nuclear energy, economic hard times set in and forecasts of future energy demand growth were scaled down drastically.

Shortly after the war, the American Atomic Energy Commission was established and assigned responsibility for future nuclear developments, civilian as well as military. At the same time, the Congressional Joint Committee on Atomic Energy was established. For the next twenty years the executive agency, and the Congressional committee, worked closely together and, in effect, reigned jointly over the governmental programs in question.

The programs in support of civilian nuclear power grew out of the programs to design and develop nuclear power reactors for submarines and surface ships. President Eisenhower's 'atoms for peace' speech in 1953 signaled, and put in place, a commitment by the United States government to develop civilian nuclear power reactors. The bulk of attention was focussed on the light water reactors for which some experience had been accumulated in the naval programs. Light water reactors needed enriched uranium as a fuel, but the United States had ample enrichment plant capacity, built in support of the nuclear weapons programs. The AEC supported R & D and demonstration plants.

It was apparent from the outset that, if nuclear power was to be competitive with conventional power, the plants would have to be very large. Thus, during the late 1950s, the companies committed themselves to produce, and utilities to buy, nuclear power plants very much larger than any that had actually been built and tested. In this era of optimism very little attention was paid to issues of reactor safety, or to the question of what to do with burned-out fuel elements.

The Shippingport demonstration plant went into operation in 1958, and was followed by the Yankee Nuclear Power Plant in 1961. Both of these plants operated at scales far smaller than the ones the companies and the utilities were already committed to produce and use commercially. The objective was to gain experience from their design, construction and use. The faith was that 'scaling up' would pose no serious problems. In 1963 a contract was signed for the first full-scale reactor, judged competitive without subsidy. A wave of orders followed.

As it turned out, the companies which contracted to build the reactors could not do so at costs anything close to the agreed price. Relatedly, there were major technical problems with the large-scale reactors that had not been apparent with the smaller demonstration versions. The first generation of commercial reactors were not competitive. The companies which produced them lost money. The utilities that procured them could undoubtedly have produced electricity at lower cost had they built up-to-date conventional plants. And this despite the heavy front-end R & D subsidy of the plants by the Atomic Energy Commission, and subsidization of fuel costs.

During the 1960s, and into the early 1970s, despite this unfortunate early experience, the companies continued to try to sell, and utilities continued to order, versions of the light water reactors. Disenchantment set in gradually. As noted, there was first a rise in concern about environmental impacts and safety, and then, somewhat later, a sharp fall in projected growth of demand for electric power. The large jump in oil prices, and more optimistic beliefs about future availability of uranium relative to demands, by themselves made the nuclear power alternative look more attractive relative to conventional plants. However, the sharp rise in estimated nuclear plant costs associated with new environmental and safety requirements, and the now much more complicated and time-consuming regulatory process, deterred many utilities from taking the nuclear route. Aside from the bringing into operation of a number of plants whose construction started some time ago, nuclear power expansion in the United States has virtually come to a dead stop.

In the early 1960s, on the belief that conventional light water reactor technology had been established, the Atomic Energy Commission shifted its own attention toward research and development on a breeder reactor. The case for the breeder reactor rested, in large part, on forecasts that there would be very considerable growth during the last decades of the twentieth century in the number of regular nuclear plants, and that supplies of uranium would therefore relatively quickly become mined out. As with the case of conventional reactors, the Atomic Energy Commission relatively early in the game committed itself to a particular type—the liquid metal fast breeder reactor. Considerable funds went into research and development on this reactor. By the middle 1970s, however, skepticism began to be strongly voiced. In the first place, projections of growing scarcity of uranium no longer seemed justified. In addition, concern that breeder reactors generated materials that could be used in bombs intensified. A number of studies argued that no economic case could be made for going ahead with at least this particular breeder reactor program. None the less, funds continued to go into the Clinch River breeder reactor project. While the old Atomic Energy Commission was dead for more than a decade, the political momentum of the projects it initiated proved hard to slow down. In late 1983 Congress stopped funding the program.

The stories of the British and French programs have some essential things in common with the American experience, and some important differences. One major difference is this: after the war both the British and the French opted for a gas-cooled graphite moderated reactor design for two central reasons. First, these reactors used natural uranium as a fuel, and their employment in a power grid therefore did not require access to enriched uranium which, in the early post-war era, only the United States could produce. Second, this kind of reactor produces plutonium as a by-product. Thus, these reactors were a natural part of a program aimed to develop a military nuclear capability. In Britain and France, as in the United States, a central government authority was established to guide and promote the development of nuclear technology for both civil and military purposes.

The British Atomic Energy Board, later the Atomic Energy Authority, has over the years exerted even more detailed control over the development of nuclear power than did the US Atomic Energy Commission. From the beginning, the lead British government agency was committed to its own particular reactor design. Electric power generation and distribution in Britain is nationalized, and centralized. The Central Electricity Generating Board was, after its early experiences with experimental plants, increasingly skeptical about the economic merits of the British design, and has over the years pressed for light water reactors. However, until recently, higher political authorities have ruled in favor of the judgments of the Atomic Energy Authority.

Britain's reactors have never found a market abroad and have only been employed domestically because the Electricity Board has been, in effect, ordered to do so. In the late 1970s and early 1980s this situation was reluctantly recognized at the top. The power of the Atomic Energy Authority to dictate the path of nuclear power development in Britain has apparently been greatly attenuated.

The French case has much in common with the British, although from the beginning the authority responsible for the nationalized power network, Électricité de France, has been a more effective counterweight to the Atomic Authority than has been the case in Britain. As it gradually became more expert, EdF became skeptical about the economics of gas-cooled graphite moderated reactors, just as had the British Central Electricity Generating Board. Until the middle 1960s in France, as in Britain, the central government authorities ruled for the atomic energy authority and against the electricity authority when there were cases of conflict. However, by the early 1970s, EdF had begun to win the upper hand, and to call the tune regarding reactor development and purchase.

By the middle 1970s France had shifted over virtually completely to pressurized water reactors as the technology of choice for the short and medium run. To a greater extent than in the United States, the designs were standardized, and the French company engaged in the production of such plants, Framatone, began to gain advantages of economies of scale and experience. While recognition that future demand for electricity will not be as great as forecasted has slowed down the pace of construction, all new electricity generating capacity in France is now nuclear, and production is planned ahead at a modest rate. France continues to work, now increasingly in consort with other European countries, on a breeder reactor.

The German story diverges from the British and French. The fact that Germany was not trying to build up a military capability is important to recognize. Also, there was no strong resistance in the German case to dependence on the Americans for fuel. Given the questions being explored in this essay, however, the most important differences were probably that strong centralized control of reactor development never took shape in Germany, nor was civil development much tangled with military programs.

After the war, for a period of time, Germany was expressly prohibited from engaging in a variety of nuclear research activities, and only in the 1950s did the constraints loosen, and the Ministry for Atomic Questions come to be formed. Historically, the Länder have had major responsibility for funding research at the universities and as Germany began to re-establish a nuclear research capability, the responsibilities were not centralized as they were in other countries. Also, like the United States, and unlike France and Britain, in Germany electricity production and distribution is not centralized—there are a number of independent utilities—and cannot be directed from the capital. The larger German companies, principally Siemens and AEG, had been watching reactor developments for some time, and when the German program got under way, had some judgments of their own as to the most promising roads to follow.

The programs of the federal government, while substantial and directive by the late 1950s, must be understood as being only a part of the action. There were a number of different sources of initiative. To a considerable extent, the major companies laid their own bets, which differed from those of the government.

By the late 1960s German companies had acquired sufficient competence to cut their ties with American firms. German reactors were competitive on world trade. Recently, falling expectations about future energy demand have led to a cutback on orders.

What lessons?

What are we to make of these histories? What do they tell us about the kinds of government R & D support programs that will work, and will not work, in different contexts? I think the lessons must be drawn delicately and cautiously. But several matters seem relatively clear, to me at least.

First, government support of generic research specifically targeted to an industry is a powerful, and apparently widely applicable, instrument. The US Department of Agriculture's support of generic research, MITI's program of generic research in semiconductors and computers, and the old NACA's programs clearly had important positive effects on the industries they aimed to help. When a government agency focuses its funds on generic research it is likely to be supporting work that industry would not do much about on its own. When industry is consulted closely regarding the allocation of such efforts, the research is likely to be well targeted. And if firms in the industry have special access to the ongoing research, they do indeed seem to gain advantages over firms that do not. In this paper I have briefly sketched a very few systems of government sponsored generic research focused on the needs of particular industries. It would seem useful to enlarge the sample, and deepen the analysis.

A second lesson is that procurement-associated research is mainly good for helping governments get what they want to procure. When that research involves a large generic element, commercial capabilities are likely to be enhanced as well; similarly, for procurement-orientated design and development, if commercial and government products are quite similar. However, procurement-related R & D is surely not cost-effective if the principal objective is to enhance commercial competence. There are important examples of 'spiller' from American defense and space R & D but the value of these is certainly puny compared with the dollars expended. It also seems to be the case that the attempt to bundle promotional and procurement objectives is a bad idea.

Expressly promotional programs are tricky, and prone to costly blunder. However, there are in our sample several examples of moderately successful ones and some lessons to be drawn. In the United States case of applied research aimed to help farmers, the farmers have had a very great influence in guiding the program. In the case of Airbus, the companies which produce the plane, and the airlines which buy it, in effect determined the designs to go after. In both cases, if for different reasons, the companies involved saw that they could not engage in the endeavor on their own, but that a cooperative effort might be profitable, particularly if the government picked up the tab. In the more egregious examples of publicly-funded promotional R & D, a government agency in effect decided what it thought ought to be done, and pushed it through. Concorde, and the American Supersonic Transport, are signal examples of what not to do. So, too, with many of the nuclear programs. Again, the sample is small, but the lessons do seem clear.

I want to conclude by returning to the theme I briefly introduced at the beginning of this essay, but which I have not elaborated on in any detail. It is that most studies of the process of technological advance in industry have tended to focus strictly on the private actors, and the studies which have been concerned with the public actors have ignored the private ones. However, in each of the cases explored here one can see private and public actors of a number of different sorts involved. Producers, consumers, universities, free-standing laboratories, government agencies, sometimes regulatory bodies, are all playing important, but different, roles. The picture is much more complex than most observers, or scholars, have painted. We will not come to understand it well, and certainly will want to be able to make good judgments about how we influence it effectively, until we learn to see this more complicated picture—or I had better say pictures. For there were striking differences, as well as similarities, among agriculture, electronics, aircraft and nuclear power. And these differences, as well as the similarities, are important to understand. But that is a story for another paper.

Notes

1. Scarcely an issue of *The Economist*, or *Science*, to mention two prominent journals reporting news regarding policy in this area, fails to mention one or more new government programs being launched, or debated.
2. His *Capitalism, Socialism and Democracy* (New York, Harper and Row, 1942) is the canonical reference. Christopher Freeman's view of industrial innovation is clearly 'Schumpeterian' in spirit. Freeman's book, *The Economics of Industrial Innovation* (London, Frances Pinter, 2nd edn., 1982) is the best available general account of technological advance, and competition, in high technology industries.
3. *Research Policy*, a journal Freeman helped to found, is the repository for many of the best of these studies. I shall give specific references when I present material on particular cases.
4. The analysis here follows along the lines I have laid out in several earlier publications, in particular, *Government and Technical Progress: A Cross Industry Analysis* (New York, Pergamon Press, 1982), and *Policies in Support of High Technology Industries* (ISPS Working Paper 1011, New Haven, Yale University, 1984).

5. There are many good studies of the programs described here. I have drawn heavily from Robert Evenson's essay, 'Technical Change in U.S. Agriculture' in Nelson, *Government and Technical Progress*, op. cit.
6. The following discussion of electronics, and that of aircraft and nuclear power, are compressed versions of the accounts given in Nelson, *Policies*, op. cit., note 4.
7. This account of United States policy towards semiconductors and computers draws heavily upon the essays by Levin, and by Katz and Phillips, in Nelson, *Government and Technical Process*, op. cit. See also Wilson, Ashton and Egan, *Innovation, Competition and Government Policy in the Semiconductor Industry*, (Lexington, Mass., Lexington Books, 1980; S. Kalos, *The Economic Impacts of Government Research and Procurement: The Semi-Conductor Experience*, unpublished Ph.D. dissertation, Yale University, 1983.
8. The following discussion is based principally upon the following sources: M. J. Peck and R. Wilson, 'Innovation, Imitation and Comparative Advantage: The Performance of Japanese Color Television Set Producers in the U.S. Market' in H. Giersch (ed.), *Emerging Technologies: Consequences for Economic Growth, Structural Change and Employment*, Tubingen, J. C. B. Mohr, 1982; M. J. Peck, 'Government Coordination of R & D in the Japanese Electronics Industry,' unpublished mimeographed essay, Yale University, 1983; T. A. Pugel *et al.*, 'Semi-Conductors and Computers: Emerging International Competitive Battlegrounds' in R. W. Moxen *et al.* (eds), *International Business Strategies in the Asia–Pacific Region*, Connecticut, JAI Press, 1983; J. Wheeler *et al.*, *Japan's Industrial Development Policies in the 1980s*, New York, Hudson Institute, 1982; and D. L. Doane, 'The Generation of New Products and Industries,' unpublished manuscript, Yale University, 1983. I am particularly indebted to Donna Doane for having made available to me her draft manuscript.
9. The following discussion of the European experience draws in particular upon Sciberris, 'The UK Semi-Conductor Industry' in K. Pavitt (ed.), *Technical Innovation and British Economic Performance*, London, Macmillan, 1980; J. Zysman, *Political Strategies for Industrial Order: State Power and Industry in France*, Berkeley, University of California Press, 1977; G. Dosi, *Technical Change and Survival: Europe's Semi-Conductor Industry*, European Research Centre, University of Sussex, 1981; and F. Malerba, *Technical Change, Market Structure and Government Policy: The Evolution of the European Semi-Conductor Industry*, unpublished Ph.D. dissertation, Yale University, 1983. I am especially indebted to Franco Malerba for helping me to understand the European record.
10. A good source is the chapter by Mowery and Rosenberg in Nelson, *Government and Technical Progress*, op. cit.
11. The following draws heavily upon R. Miller and D. Sawyers, *The Technical Development of Modern Aircraft*, London, Routledge and Kegan Paul, 1968; G. Eads and R. Nelson,'Government Support of Advanced Civilian Technology: Power Reactors and the Supersonic Transport,' *Public Policy*, (1971), J. Newhouse, *The Sporty Game*, New York, Alfred Knopf, 1982, and many issues of *The Economist*. I am indebted to William Spitz for helping to pull this material together.
12. The most important sources of the following discussion are: Eads and Nelson, op. cit.; W. Walker and M. Lönnroth, *Nuclear Power Struggles: Industrial*

Competition and Proliferation Control, London, Allen and Unwin, 1983; O. Keck, *Policymaking in a Nuclear Program*, Lexington, Mass., Lexington Books, 1981; and G. Hazelrigg and E. Roth, *Windows for Innovation: A Story of Two Large Scale Technologies*, submitted to the NSF by Econ, Inc., 1983. Michael Sullivan ably surveyed the European experience.

PART II

Science and Technology Policy

4 Science in the political arena

Jean-Jacques Salomon

The idea of knowledge as something that should illumine and guide the exercise of power is as old as political thought in the West: the one appeared with the other. 'Knowledge' and 'power' are opposites; their conjunction must bring about a better, if not the best, government of society. From the outset in political thinking, this hope has been present as a basic postulate of Reason. Knowledge on the one hand, strength on the other—is it possible to unite the two in a single person? The reply to the question given by Plato is the oldest *aporia* in political thought: it will suffice 'for philosophers to become kings or else for those that are at present called kings or rulers to become genuine and competent philosophers'.[1]

The ideal still survives that the right knowledge can ensure sound political action: beyond Platonic idealism, there is the notion which informs all Western rationalism, that understanding permits the mastery of its object. However, the ends to which the Greeks applied their understanding are utterly unlike those that concern us—besides which, neither knowledge nor time had the same meaning for Plato that it does for us. If progress in understanding was possible for the Greeks, progress in time was inconceivable: whether decline in the sense of biological deterioration or repetition in the sense of an eternal recurrence, the cyclical nature of time was part of an order of things over which men had no control. What the philosopher taught the statesman was the best way to reach an understanding of this order, so that the statesman would thereby have the greatest chance of ensuring just rule in society. The knowledge on which the philosopher drew was a virtue, not a power.[2]

Since Plato there has been no lack of philosophers to play, with greater or lesser success, the role of counsellors to the prince, attempting to inspire political action with their own brand of philosophy. The species has continued to proliferate, first with Christianity, then with the scientific revolution of the seventeenth century, and most of all with the industrial revolution. But, just as the possibility of progress in time came to be perceived, so the nature of knowledge changed, and it is precisely the coming together of the transformation of knowledge and the rise of an historical consciousness that has defined 'modernity'. From Saint

Augustine to Marx, via Vico, Hegel and Comte, the philosopher's advice to the statesman has no longer been concerned with contemplating the cosmos, but rather with changing the world.

It is, of course, not by chance that the ultimate avatar of modernity can be seen in the influence exerted today by the scientist on the political scene: science is one of the most effective ways of changing the world, and hence the dream of a fusion of knowledge and power has in some ways been realized. But how far does it go—and at what cost? If the paradoxical relationship between the two partners has altered, it is in style rather than substance: the trick of Reason has simply laid new traps for them.

Few contemporary economists have been more aware than Christopher Freeman of these traps. Disappointed by the arbitrary nature of political theory, the former student of Harold Laski at LSE turned instead to the more empirical study of the economics of research and development and of technical change; rather than getting involved in a purely ideological debate, he preferred to try to understand the forces that determine social change. But the economist has never forgotten the teaching of Plato: mastery of a subject can help to guide the practice of it, provided that the intellectual approach is not separated from an awareness of the values involved.

This chapter is contributed in homage to the political mind that has always remained hidden, subtly but firmly, inside the economist in Chris Freeman. I offer as proof the following passage from a lecture that he gave more than twenty years ago, during a seminar that I organized—a lecture that had a considerable influence on the subsequent development of thinking as regards science and technology policies in the member countries of OECD:

> I certainly would not claim that an economist, just because he is an economist, has any right to determine these political, strategic priorities. They must be the outcome of the political process. This is not to say that the political process could not be very much improved; undoubtedly it could be improved and I shall make various suggestions by which it might be improved. Although the economist's vote counts only as one along with all other citizens in determining the main political objectives of society, he does nevertheless have a special responsibility in contributing to the discussion formulating these objectives.[3]

I

The modern bond between the scholar and the politician arose out of the divorce which separated science and philosophy. The scientific revolution personified by Galileo, Descartes and Newton provoked the breach which has grown progressively wider with the developing professionalization and specialization of a new intellectual community, that of the natural philosophers who, from about 1840 onwards, have gradually come to be called 'scientists'. Bacon's famous dictum: 'Knowledge is power' sums up the whole upheaval generated by the scientific revolution which replaced a contemplative science that separated theory and practice with one that is entirely geared towards the practical application of ideas, by reducing nature to mathematical terms and making its proofs through experimentation.

The savant is no longer someone who seeks virtue, or justice or wisdom, nor does he devote himself entirely to theoretical reflection; rather, he is someone who possesses a practical understanding that allows him to manipulate the physical world, and ultimately men too. And unlike philosophy, which almost by definition is directed towards challenging the order of government, this new science that generates practical knowledge is by nature closely linked with political power; whatever the regime, science offers its service to the State and in return it expects the State to support its research. The political element in science was built into its intellectual foundations: its experimental character presupposes instruments, special sites, financial resources, and therefore the protection of the prince, who will provide these things all the more willingly because science promises to deliver useful results.

Science certainly promised the moon, but it was still a long way from being able to fulfil its promises. Apart from extreme cases, usually generated by war—the French Revolution with 'les savants de l'An II', the War of Secession, the Franco-Prussian war—science by and large was left alone, because it did not have much to offer in terms of practical applications and at the same time incurred little expense. Government without science was indeed still possible until after the First World War. And yet, from the very beginning of this alliance, the first snare was there waiting for the two partners: the connection between the scientific establishment and the political environment, which began with the creation of scientific societies and the growth of the first laboratories, contained the inherent risk that scientific progress would be subjected to the requirements of the State.

Science, at the same time that it promises the prince useful results and consequently expects his support, also requires his complete neutrality with regard to its discourse and its methods. The practice of science, as the Royal Society's charter lays down, should not involve 'meddling with politics'. Politics are to be excluded in so far as they represent an opinion or position that might interfere with the procedures appropriate to scientific research; and yet, politics are implicit as the point of contact of the two interest groups, researchers and the State. The prohibition against politics encroaching upon the intellectual process does not prevent there being an accord with government; but the accord contains in embryo a threat whose reality has been evident from the very outset of modern science, as the Galileo affair made clear.

Political power by definition means authority, and nothing seems more hostile to the scientific process than to have to take account of any authority other than its own intellectual requirements. As has been noted, political power attributes authoritative values:[4] so how can science be subject to the values of power and still remain true to itself? In this respect, the history of science is bound up with the history of its conflict with the spirit of authority, whether embodied in the dogmas of the Church, the official doctrines of the State or the orthodoxy of some party.

I know of no text that defines more clearly the limits to be fixed between the areas of competence of the scholar and the politician than the letter from Galileo to Christina, Grand Duchess of Tuscany. These limits have not been in any way removed by the ever closer association today between science and the State; such an intervention in the work of scientists, Galileo says, 'would amount to commanding them that they must not see what they see and must not understand what they know, and in searching they must find the opposite of what they actually encounter'. The truths which are the province of science cannot be refuted by any other authority:

> With regard to these propositions and others like them which are not directly matters of faith, certainly no man doubts that the Supreme Pontiff has always the absolute power to approve or condemn; but it is not within the power of any created being to make things true or false, for this belongs to their own nature and to the fact.[5]

One may question the autonomy of science, as happened in the aftermath of the upheavals of 1968, and one may show that the internal logic which dictates scientific progress is nevertheless also linked to economic and social change; one may criticize the claims of an exclusively endogenous approach to the history of science and take account of the social origins and conditioning that have shaped the scientific institution, its members and its products; one can again force laboratories to do more to meet the needs of society, as was done recently in France under Giscard's policy of '*pilotage par l'aval*' (moving with the tide) or under the Socialist government's appeal to 'social demand'. But the idea of the autonomy of science—even if the current evolution in research activities, the mingling of science and technology, the ever closer links between university laboratories, industry and the State, and the influence of the military-industrial complex on the nature of the research undertaken all justify protests about its limitations, as well as its ideological mystifications—this idea of autonomy relates to an incontestable reality: that is, first and foremost, the acknowledgement of the rights of truth against any authority which seeks to diminish or falsify it, or subject it to the fiats of some orthodoxy.

The Nazis impugned Einstein's theories in the name of 'Aryan' mathematics and physics, the Communists challenged Mendel and the theories of genetics in the name of dialectical materialism. In the two cases the results were the same: Germany under Hitler and the USSR under Stalin both deprived themselves of the knowledge and the methods which the theories under attack provided towards helping the progress of science elsewhere. The State, the party or the dogma that tries to dictate to science the rules governing its activities, to arbitrate in the controversies which concern the scientific community, or to exclude from that community those researchers who do not comply with the orthodoxy in power, thereby exhibits the very essence of totalitarianism.

This aspect of relations between the scientist and the politician provides plenty of examples—including victims who paid the price with their lives—to give strength to the arguments of scientists and partisans of *laissez-faire* whether from intellectual or political bias. However, as they have moved from the setting-up of scientific societies to the environment of universities, as they have developed and multiplied as an increasingly professionalized community, academic scientists have always favoured a one-sided version of *laissez-faire*: the State should support research, but without the counterpart, that is, without any right to oversee the internal affairs of science. If the pursuit of knowledge is a good in itself, whose long-term spin-offs will be beneficial to society, society should leave well alone and should not demand some right of intervention over its conduct.

From the nineteenth to the twentieth centuries—from Renan to Polanyi—the idea of the autonomy of science involved an implicit double standard: support for research is the State's duty, a duty which arises from the collective interest, but the contract that binds the State lays down no reciprocal obligations for science.[6] On the contrary, the less the State imposes itself in this area, the better the chance science has to move forward: the slightest intervention brings with it the threat of excessive control, the possibility of the substitution of a new order of values, the encroachment of the authority of violence or bureaucracy on that of truth. One has only to read *The Third Circle* by Solzhenitsyn or *Life and Fate* by Vassili Grossman to recognize that the meddling of the State or the party in the domain of basic research always carries with it this risk of forcing a wrong turning.

II

Polyani was writing his warning just after the Second World War, at the time when there first was talk of formulating and implementing science policies, that is, with reference to the mechanisms created and the measures taken by governments in all the industrialized societies simultaneously to develop scientific activities and to derive the benefits of the results of research within the framework of national objectives; the era of *laissez-faire* was already over. The success attributable to science during the war laid a new trap for the relations between science and power which none of the founding fathers of modern science (even Bacon) could have envisaged.

Policy for science, but also policy through science: the source of the trap lay in the fact that government and science were henceforth inextricably linked. The two policies could not be pursued independently and both had become part of the responsibility of the State. It was no longer a matter of simply promoting scientific activities for cultural reasons; it was important henceforth to exploit the results of scientific research as one more means of political action: science was to be treated as a tool, if not a commodity, like any other. The immense research effort undertaken during the war— in particular the Manhattan District Project dedicated to the acquisition of the first atomic weapons, but also the work done on radar, operational research, guidance systems, computers, DDT or penicillin—led to the recognition of science as a 'national asset' which could determine the power relationships between nations and could contribute to the pursuit of their goals, especially in the military and economic fields.

However, this new asset required enormous investments: the scale of the investments now needed for scientific and technological research, representing today 2 to 3 per cent of gross national product in the most industrialized countries, is such that the scientists have no alternative but to accept the support provided by the State. And, given the absence of peace with which the war ended, the competition between the Great Powers and the technological race which resulted from it, the rate of growth of this commitment, often faster than that of national product, absorbs an increasing proportion of public spending. Henceforth, the government could no longer leave science to itself: the scientific establishment had proved itself so profitable in political terms in wartime that it had to be mobilized in peacetime, too. Governments had to equip themselves with new agencies to coordinate and to give direction to the research effort in science and technology, surrounding themselves with scientific advisers and taking measures dedicated to forcing the pace of discovery and innovation. Oddly enough, it was America—champion of the free market economy—that provided the model of political intervention in technical and scientific matters, even if now that same country nags its most advanced Western partners, especially Japan, for following the very 'targeting policies' in which she led the way.[7]

The first consequence of this development was that the one-sided contract with the State was revealed to be unworkable; the alliance (that is, the common interest) of the two parties required a reciprocal obligation. It was the stimulus of military considerations and economic competitiveness that made the research establishment so indispensable to the working of politics; the contract necessarily became a matter of give and take. In spite of the upheavals within the scientific establishment, its growing industrialization and even militarization, the representatives of academic science constantly protested against any reciprocity of obligation, even as they became ever more subject to political control.

The second consequence was that the potential conflict between the scholar-scientist and the politician was no longer limited to a confrontation between truth and its opponents. The State, it is true, could not tell researchers how they should work nor, *a fortiori*, what they should discover. But it did expect them to come up with results of some kind—as in the famous words of de Gaulle, welcoming one of his ministers (a politician) responsible for scientific affairs, 'Well, my researchers, what have you all been finding?' And, indeed, even in the more *laissez-faire* regimes, the government was less and less able to hold back from indicating to scientists what research they must undertake, for instance by directing resources and hence scientific personnel into one area or discipline rather than another; apart from military research, there is no better illustration of this trend than space information technology and biotechnology. It is characteristic of the new relationship between the scientist and the politician that the confrontation between them is now not simply on the grounds of truth, but also on those of results, productivity and profitability.

The upholders of *laissez-faire* will argue that these grounds are not appropriate ones on which basic research should be called to account, or the academic scientists will claim that what is relevant in terms of policy-making for technology does not apply to science. But in fact both the times and institutions have changed: the scientific community is no longer made up of a handful of savants whose special setting is the university laboratory; there are now hundreds of thousands of researchers, and those who devote themselves to pure science are only a tiny fraction of the scientific work-force, among the vast mass of researchers, technologists, engineers and technicians without whom the development of science itself would now be impossible. The greatest number of scientists as such is nowadays to be found not in academia, but in industry and government laboratories. It is not an accident that the word used for the researcher has changed: the savant was part of a culture where research was craft-based, whereas the 'scientist' is defined in terms of an occupation, a career, a status within a professional, technically skilled group. The 'man of science' relates to a vocation, whereas the 'scientist' suggests a specialist in a job like any other—hence just one productive element among the many created by industrialization.

Above and beyond this, research activities today are part of a *continuum* in which it is difficult to identify what is basic research and what is not. A scientist can go about his research like an engineer, or an engineer like a scientist, and the motivations of the one as against the other are not sufficiently distinct to delineate the boundary between 'pure' science and applied research. A laboratory in industry can undertake fundamental research under the same conditions as a university laboratory. Neither the motivations of the researcher nor the substance of his research allows the distinction to be made between 'pure' and 'applied'. At best the demarcation line can be drawn on the basis of spatio-temporal factors: the *environment* (an academic setting rather than a government laboratory or private industry) determines more or less the *time* factor, and hence the degree of *freedom* relative to the goals pursued. The measurement of time, too, is a function of the conditions—the pressures—which prevail in a given situation, and even the university is not sheltered from these pressures.[8]

This does not mean that research where results are expected only in the medium or long term or are simply not foreseeable does not play an essential part in the whole system: basic research is central to scientific education and to the advancement of knowledge and of our general culture. In the true sense of the term, it is a kind of insurance for the future, because it is one of the sources of both specifically technical innovations for tomorrow and of broader concepts or theories which will transform our vision of ourselves and of our world. Yet is it possible to set apart from the whole research system those kinds of scientific research that are out of reach of all demands and pressures exerted by society? And how do we make choices in an area where all the disciplines are competing for special support towards similar ends? Ultimately, even the academic scientists whose work is furthest from any economic considerations are dependent on the public purse to finance their laboratories.

It is hardly surprising, in this situation, that even 'pure' science is a matter for political debate, decisions and directives. One can understand the nostalgia, on the part of scientists who were trained before the Second World War, for the 'golden age' when the scientist could still proclaim his 'independence'. This nostalgia has even been expressed in the Soviet Union when Kapitza, paying homage to Rutherford, his teacher at Cambridge, declared,

> The year that Rutherford died, there disappeared for ever the happy days of free scientific work which gave us such delight in our youth. Science has lost her freedom. Science has become a productive force.

> She has become rich but she has become enslaved and part of her is veiled in secrecy.[9]

This nostalgia lingers on, but it is not simply a preoccupation (or an ideology) of an earlier generation—what is at stake goes much deeper. There is, in fact, a close connection between the idea of 'pure' science and the Platonic conception of knowledge. The nostalgia is for the world which could be grasped 'purely' in terms of Ideas. This can be seen clearly in the recent book by René Thom, the great mathematician and winner of the Fields Medal, who pleads for a scientific approach unlike the one that we are familiar with: by the same token whereby he looks forward (or backward?) to a science which would be more aware of 'form' than 'force', he criticizes the scientific establishment for being too close to government and to technical-industrial preoccupations. Just as the devotion of physicists to the worship of matter and force has caused them to neglect the epistemological importance of form, so in Thom's eyes the growth of experiments aimed at useful results has led to ignorance of the 'real' scientific implications.

> Recently [Thom says], I heard a colleague in the Academy praising the merits of his own discipline in terms of the business deals made in France by the corresponding industrial sectors. Such an argument, in my own view, is totally monstrous and makes me want a much clearer segregation between science proper and its technological applications'.[10]

By turning away from the purely speculative conception of Greek science, modern science has gained social legitimacy and political weight, as can readily be proved. Are then the scientists who practise it like the prisoners in the Cave who see only the shadows of the Ideas? It is nevertheless these shadows that today make the world work. 'Segregation' has become just as impossible as the contract without any reciprocal obligation. The quantitative change which has affected research activities since the end of the last war has been accompanied by a qualitative change, and the rules of the game no longer fit in with the 'ideal' standards of the scientific craftsman: they are affected by the same shortcomings and malfunctioning that accompanied the whole process of industrialization. The American phrase 'publish or perish' sums up perfectly the notion of this Darwinian struggle for survival, because the rewards that publications or results can bring in prestige will also guarantee renewal of support—*credits* in both its meanings. One has only to read *The Double Helix*, in which James Watson tells the story of his winning the Nobel prize for medicine and physiology, or Nicholas Wade's account of how Roger Guillemin and Andrew Schally won the same prize, to see how little the standards of conduct in the university laboratory resemble the behaviour which seems to have been characteristic of earlier generations of scientists.[11]

No wonder if, in the view of certain academic scientists, this change seems like a betrayal of the goals and the interests of science: scientific research should only concern itself with the advancement of knowledge, remote from the pressures, the temptations and the conflicts of values which sustain the world of politics. The perspective of utility in which science has flourished has also compromised it, perverted it and, to be explicit, prostituted it. But the State inevitably has an instrumental view of research, and the rules which govern the support it gives are shaped by the goals which it adopts. The support of all science, even the most basic, is now subject to bureaucratic processes of evaluation and control; the scientists are no longer left to be accountable in their work to their peers alone.

III

Just as the scientist is in a position of dependence with respect to the State, so the modern State is itself dependent upon the scientist. Plato's kings or rulers, whatever the good advice from the philosophers whom they chose to listen to, could quite easily do without it and nevertheless govern successfully. Now, however, no State can survive without the advice, the help or the contributions of scientists.

It is true that it has taken much longer for the American scientific community than for its European counterparts to complete its apprenticeship in the political arena. 'The scientist going like Mr Smith to Washington' is nowadays a well-documented species of the genus political animal; before the war, the species was almost unknown in the United States, while in Europe it has been breeding ever since the 1920s. The Old World already had a long experience of dialogue between scientists and politicians, even if, after the Second World War, it was in the New that it sought its inspiration as regards science policy (imitating in particular the model of organization of OST, the Office of Science and Technology, and PSAC, the Presidential Scientific Advisory Council). And it is not accidental that, along the road to Washington, a good part of the journey was first made by immigrant European scientists (Szilard, Bohr, Fermi, etc.), less embarrassed than their American colleagues at the prospect of talking to congressmen, statesmen, members of the military or bureaucrats from the Administration.

The fact is that the inter-war experience was very different in Europe, marked as it was by both the implications of the ideological and political debate (scientists took positions with respect to Nazism and communism) and the example of the Russian Revolution (where science, defined as a public service, was claimed to be 'integrated' into the social system and planned as a productive force along with the rest). While *laissez-faire* remained an article of faith in the United States, it was more easily called into question in Europe, in particular by left-wing scientists who wanted far greater state intervention in support of science (Bernal in England, Perrin in France). Thus, as early as 1936, France was the first non-communist country in which the government took responsibility for basic research.

The political situation explains why, well before the United States, France should have been prepared to undertake a deliberate formulation of science policy. But the institutional, social and educational context also explains why the dialogue between scientists and politicians was so much easier. As is well known, it took nothing less than the famous letter from Einstein to draw Roosevelt's attention to the military importance of nuclear power (the letter was actually writen by Szilard—it took a Nobel prizewinner, and that particular one, to act as a mailbox between the scientific milieux and Washington). In France, the links between researchers and the administration were facilitated by the fact that both were products of the scientific Grandes Écoles: a graduate of the École Polytechnique who holds an important position in a ministry (sometimes even the minister himself) speaks the same language as a scientist. Raoul Dautry, Minister of Armaments in 1939, had no trouble in understanding what was at issue in artificial radioactivity when Frédéric Joliot-Curie filed with the CNRS the patents which would cover both civilian and military purposes. The alliance between science and politics went through a long period of engagement in the United States; in France they were aready cohabiting.[12]

More importantly, one must bear in mind the American doctrine—both ideological and constitutional—of non-intervention by the federal government in matters of education and research, which broadened the economic and political notion of the free market in which the State should intervene as little as possible. The Old World, where most countries already had centralized governments and public systems of education and research directly funded by the State, was accustomed to a tradition of economic interventionism which even the return from time to time of more free-market style regimes never entirely eliminated. France, with its tradition handed down from Colbert, could be seen as an extreme case of centralization and of the *dirigiste* approach, but many other European countries showed the same tendencies.

There is obviously a close link between the European pattern of the 'mixed' economy and the collusive relationship which developed in Europe between science and government before the Second World War. Whereas, after Vannevar Bush's *Science the Endless Frontier*, it took five years for the American Congress to recognize (by creating the National Science Foundation) the necessity for the government to take charge of basic research, that development was self-evident in Europe. And then it did not happen, as it did in the United States, in the name of an economic doctrine whereby science, if 'abandoned to market forces', would be condemned to the smallest share,but rather it was the result of political considerations which conceived of science as being just as legitimately an area of state concern as any other.

Because of the strategic character which science would have henceforth, the American system was obliged in its turn to recognize the inevitable and irreversible conjunction of public and private interests: 'the blurring of the boundaries between business and government was the earlier blurring of the boundaries among the sciences, and between the sciences and engineering; but the more urgent reason was the need to advance weapons technology in the interest of national defense'.[13] This 'fusion of economic and political power' for which scientific matters have been the alibi none the less remains heretical in the United States. Whereas, to quote Don K. Price again, the federal government has had to learn 'how to socialize without ownership', the majority of European countries were already quite used to such a socialization, through their experience of nationalizations on the one hand and through the existence of a large public sector in higher education on the other.[14]

The exercise of power is today so closely connected with scientific resources that there is no policy, foreign or domestic, which does not depend on the process, the methods, the results and even the promises of scientific research. This is not to say that the art of politics has become more scientific now because it has become more dependent upon scientific methods and tools. Scientific input is a necessary but far from sufficient condition for policymaking, and this is where the third trap lies in wait for the hope or the mirage of a 'fusion' of the two opposites in the old paradox.

One of the characteristics of the modern State, according to Weber, is the growth of a trained class of civil servants. Modern bureaucracy embodies knowledge: 'Technical knowledge is required for the discharge of administrative tasks due to the modern technology and economy of production, and it is irrelevant whether production is organised in a capitalist or socialist fashion'.[15] Weber clearly saw that, whatever the regime, technical competence is involved in efficient management. He did not realize, however, that scientists in their turn would become actors on not only the bureaucratic, but on the political stage as well.

As Jurgen Schmandt has shown, there are at least three levels where science is now involved in the administrative and policymaking process:

as product, as evidence and as method. The scientific State is 'the attempt to merge legal with scientific rationality'.[16] In other words, the modern State is bureaucracy with science added. Schmandt concludes from this that 'the administrative state has lost the rigidity which resulted from exclusive reliance on legal rationality, even when the increased technical complexity of modern life was no longer adequately dealt with by this principle'.[17] But will this increase in flexibility, by reinforcing and increasing the strength of political power, bring it to the point where it is all-powerful? This broadening of legal rationality by scientific rationality contains all the Orwellian visions and threats of *1984*. Nevertheless, there are two good reasons why reliance on this new principle of the scientific State does not lead to some nightmarish situation: first, the construction of scientific evidence is, after all, just another social process; second, such a construction is basically altered as soon as it has to take into account the ambiguous conditions of the policymaking process. In the political arena, in brief, the logic based on scientific evidence neither supersedes nor replaces the logic based on convictions.

There is at the same time a new paradox in the relations between scientists and government. Politicians are rather more accustomed than are the scientists to dealing with ambiguous situations, and yet they expect that scientific method will get rid of uncertainty and ambiguity. From the social sciences to the 'hardest' sciences, all the controversies of the last quarter-century involving reliance upon scientific rationality have shown how disappointed policymakers must be in this hope: once science has to make pronouncements on political issues, there are inevitably uncertainties in the evidence supplied by the experts, who tend to reveal their differences instead of presenting a united front. The evidence or the proofs which the scientists provide for the policymaker can at best reduce uncertainty, they cannot remove it entirely. The facts that science can provide in its own field are not equally solid; and even when they are solid, they lose that solidity the minute that they become involved in political debate.[18]

Moreover, in spite of the expansion of bureaucracy and the State, it cannot be said that there has been any growth in the rationality of control in public affairs; in spite also of the growing input of scientific knowledge and methods into policymaking, it cannot be said that we are moving towards a more rational political society. It is true that the complexity of modern societies sets limits on the effective democratic control by the citizen-body, which supports Habermas's thesis that the really important decisions are made in the administrative and political hierarchies by means over which individuals and the citizen-body are increasingly aware that they have little or no control. However, the growing autonomy of the political system in Habermas's sense is one thing and the extension of rationality is quite another.[19]

For example, in those societies where there is public debate and where opposing points of view are expressed, technical expertise is not a substitute for the normal system of checks and balances. Even in closed, totalitarian societies, it can hardly be said that rulers rely any more on the principle of scientific rationality than on the traditional techniques of dictatorship and tyrannical bureaucracy. No 'fusion' is implied between knowledge and power which would suppress the accidents of history or men's conflicting passions and values. In a sense, rather than undermining the political prerogative, the scientific management of modern societies leads to the undermining of the image of science and of its prerogatives, naïvely thought to be the universal key to the problems of governing society.

In addition, the State now has to deal with and decide upon many questions raised by the scientists themselves. Without Einstein there would have been no Manhattan Project, because he was not only the man who will go down in the history of science as the discoverer of the equation linking energy and matter, but also because he will go down in the history of the world as the man who persuaded Roosevelt of the strategic importance of atomic research. Scientists are, in fact, the only technicians who can manipulate nature itself, propose ways of altering its state and conditions, shape the knowledge and the endproducts whose spread may transform the issues and terms of the political process.

The most dramatic example, obviously, is that of nuclear weapons, where the rate of technological progress constantly changes the basis of the negotiations on the control of armaments: the definition of parity between the Great Powers or the vulnerability of a weapons system is determined by experts. But at the same time the pace of negotiations—granted that agreement is the collective goal—is never fast enough to catch up with the pace of scientific and technological innovations, as new weapons systems are created. The result is a bitter pill for the believers in arms control to swallow: technological advances constantly threaten the very security that they were supposed to assure in the first place.

In fact science has become so complex and so esoteric in many other fields besides defence that the policymakers are often at the mercy of experts. The major technological programmes—nuclear reactors, missiles, telecommunications satellites, particle accelerators, but also less costly projects which nevertheless have enormous consequences, such as those involving electronics, robotics and biotechnology—are conceived of in terms which the policymakers can only understand if they are translated into simplified language. In these fields it is clearly essential that decisions be taken at the right moment, since there are often lags of ten or even fifteen years before proposed programmes reach development stage and industrial exploitation. In that case the government can be seen to be vulnerable to pressures from representatives of the scientific community acting as a lobby like any other, or it may be condemned to taking on projects (whether on the national or the international level) for which it is impossible to forecast, let alone control, the financial implications.

The truth of the matter is that, in spite of (and often because of) being scientists, the new counsellors to the prince are far from infallible. And the technical clothing in which they parade their advice may be all the more dangerous because it lends the authority of objective judgement. Experience has shown that one cannot attribute to them greater skill in the art of political navigation than those for whom it is their profession, even if one has to admit that a sense of the long term is rarely a feature of the politician. Some of the 'atomic scientists', after the Second World War, claimed that, for having developed nuclear weapons, they should be recognized as having a special claim to authority in the conduct of affairs on the basis of the culture, methods, objectivity and the norms of science. We know that someone trained at the École Polytechnique should be competent, as the name suggests, in a variety of techniques, but it is not just an aptitude for mathematics that will particularly qualify him to play a role in world affairs. There are, on the contrary, many reasons to attribute to the scientists *naïvetés* with regard to international relations, as Warner B. Schilling has noticed. These arise directly from the scientist's education and culture: they consist in looking for a simplistic and mechanistic solution to international relations, or in postulating that a problem requires a radical leap, or in assuming that the answer to the question lies in an exhaustive study of all its aspects where in fact small advances, progressive improvement, trial and error, patience and the old formulae of diplomacy or Machiavellianism will succeed more often than rationality.[20]

This does not stop an increasing number of scientists from becoming directly involved in the affairs of state; they take part as advisers, administrators, managers, ministers, strategists, diplomats, soldiers, spies. From this perspective we are a long way from the conclusions reached by Max Weber in his famous lectures on 'Science and Politics'.[21] The scholar, without being prophet or demagogue, may live 'for' as well as 'off' politics. Such collusion between science and politics does have one important consequence: it at least disposes forever of the idealized image of science as something set apart from the ambiguities of the political arena by the neutrality of its discourse. The roles are not reversed, but they may be mingled in a troubled relationship where the scientist can no longer maintain a separation between his ideas and the use made of them. The relationship is almost analogous to the wound and the knife, as the anecdote told by Dean Acheson, Secretary of State under Truman, shows: 'I accompanied Oppie [Oppenheimer] into Truman's office once. Oppie was wringing his hands and said: "I have blood on my hands". "Don't ever bring that damn fool in here again", Truman told me afterwards. "He didn't set that bomb off. I did. This kind of snivelling makes me sick".'[22] It is not certain that by shouldering his responsibilities as statesman so resolutely, Truman thereby relieved Oppenheimer of his.

The discovery of social responsibility is the price paid for the very close connection that has developed between science and power. The discovery was unpleasant, like any cause of guilt and bad conscience, and all the more so because scientists continue to claim neutrality in their social behaviour for their discourse and their methods. As Oppenheimer wrote, 'In a sort of brutal significance, which no vulgarity, no pleasantry, no exaggeration can wholly abolish, the physicists have known sin, and that is a knowledge which they can never lose.'[23] We know now that this awareness of sin is not limited to atomic physicists. Moreover, concern with the public accountability of research has led to controversies which cut the expert's idol of objectivity down to size and bring right out into the open the commitment of certain scientists as militants like any other partisans.

For instance, the ecologists and the anti-nuclear campaign responded to governmental experts by calling in their own specialists to provide counter-evidence, which undermined still further the traditional image of science's neutrality. D. Nelkin and M. Pollak put it very well:

> When scientists expose the conflicting technical views within the scientific community, they raise public doubts about the neutrality and independence of science; yet, by engaging in a debate, they too seek credibility as a source of expertise. When scientists use their expertise to bring legitimacy to the nuclear opposition, they are working with a movement partly based on mistrust of precisely the expertise they themselves represent.[24]

Ambiguous scientific advice, ambivalent scientific loyalties: the last trap of the collusive relationship is that science has become an ordinary political issue.

On the other hand, the exercise of power nowadays requires that decisionmakers, administrators and politicians understand what science is about—not its most refined techniques or theories, but at least its general functioning, methods, results and their implications. It is not essential to be a scientist to govern, and it is equally important to be ready to follow up their advice in good time and yet not to be entirely at the mercy of their expertise. But if the statesman has at his disposal some minimum of scientific culture, that alone will not make him more able to make policy decisions—nor necessarily the right ones. From this angle, neither the achievements of science since ancient Greece nor the role that science now plays in the political arena has changed in the slightest the old Platonic question: What good is there in knowing about navigation if you do not know where to sail? President Weizmann was an excellent chemist, President Carter was trained as a nuclear engineer, M. Giscard d'Éstaing was educated at the École Polytechnique, Mrs Thatcher has a chemistry degree: the question remains open as to whether, in order to govern society *well* today, it is better henceforth that the kings or rulers should be, rather than the 'genuine and competent philosophers' that Plato dreamed of, scientists, generals, lawyers, professors, laymen, managers, or movie actors.

Notes

1. Plato, *Republic*, 473d.
2. On the Greek conception of history and the transition to the modern view, see the article by Kostas Papaioannou in *Diogenes*, **25** (January–March 1959), reprinted in the admirable series of essays, *La Consécration de l'histoire*, Paris, Éditions Champs Libre, 1983.
3. Christopher Freeman, 'Science and the Economy at the Level of the Firm', in J.-J. Salomon (ed.), *Problems of Science Policy*, Paris, OECD, 1968, pp. 56–7.
4. David Easton, *The Political System. An Inquiry into the State of Political Science*, New York, Knopf, 1960, p. 40.
5. Galileo, 'Letter to Christine of Lorraine' (1615), translated by Stillman Drake in *Discoveries and Opinions of Galileo*, Garden City, New York, Doubleday Anchor, 1957, pp. 193, 210.
6. Ernest Renan, *L'Avenir de la science*, Paris, Calmann-Levy, 1890; and Michael Polanyi, *The Logic of Liberty*, London, Routledge and Kegan Paul, 1951.
7. A revealing example of this current concern with 'targeting policies' appears in *Commercial Biotechnology, an International Analysis*, Washington, Office of Technology Assessment, Congress of the United States, 1984, pp. 475–85; see also Michael Gibbons and Björn Wittlock (eds.), *Science as a Commodity, Threats to the Open Community of Scholars*, Harlow, Longman, 1985.
8. See Harvey Brooks, 'Applied Research, Definitions, Concepts and Themes' in *Applied Science and Technological Progress*, Washington, National Academy of Sciences, 1967, p. 23; and J.-J. Salomon *et al.*, *The Research System*, Paris, OECD, 1972.
9. P. L. Kapitza, 'Address to the Royal Society in Honour of Lord Rutherford, 17 May 1966', *Nature*, **210** (21 May 1966), pp. 782–3.
10. René Thom, *Paraboles et Catastrophes*, conversations with G. Giorello and S. Morino, Milan, Il Saggiatore, 1980; rev. edn, Paris, Flammarion, 1983.
11. James D. Watson, *The Double Helix*, London, Weidenfeld and Nicolson, 1968; rev. edn, Harmondsworth, Penguin, 1970; and Nicholas Wade, *The Nobel Duel*, New York, Anchor Press/Doubleday, 1981.
12. See Spencer R. Weart, *Scientists in Power*, Cambridge, Mass., Harvard University Press, 1979.
13. Don K. Price, *The Scientific Estate*, Cambridge, Mass., Belknap Press, 1965, p. 39.

14. Ibid., p. 43.
15. Max Weber, *Wirtschaft und Gesellschaft*, (Tübingen, Mohr, 5th edn, 1976, p. 128, translated by Jurgen Schmandt in 'Toward a Theory of the Modern State—Administrative versus Scientific State', in J. S. Szyliowicz (ed.), *Technology and International Affairs*, New York, Praeger, 1981.
16. Jurgen Schmandt, ibid., p. 91.
17. Ibid., p. 92.
18. See Dorothy Nelkin (ed.). *Controversy: Politics of Technical Decisions*, London, Sage, 1979; Guild Nichols, *Technology on Trial*, Paris, OECD, 1979; and J.-J. Salomon, *Prométhée empêtré, La résistance au changement technique*, Paris, Pergamon, 1982; new edn, Anthropos, 1984.
19. Jurgen Habermas, *Legitimation Crisis*, Boston, Beacon Press, 1975.
20. Warner B. Schilling, 'Scientists, Foreign Policy, Politics', in R. Gilpin and C. Wright (eds), *Scientists and National Policy Making*, New York, Columbia University Press, 1964, pp. 154–5.
21. Max Weber, 'Science as a Vocation' in H. H. Gerth and C. Wright Mills (eds), *Essays in Sociology*, New York, Oxford University Press, 1970; on the implications of this change, see my *Science and Politics*, Cambridge, Mass., MIT Press, 1973, Part III.
22. *New York Times*, 11 October 1969.
23. J. R. Oppenheimer, 'Physics in the Contemporary World', *Bulletin of the Atomic Scientists*, **4**, no. 3 (March 1948), p. 66.
24. Dorothy Nelkin and Michael Pollak, *The Atom Besieged: Extraparliamentary Dissent in France and Germany*, Cambridge, Mass., MIT Press, 1981, p. 100.

5 The role of quantitative information in science and technology policy formulation

Charles Falk

Introduction

The development of science and technology (S & T) policy, like the development of any other policy, involves complex processes which are affected by numerous internal and external factors.[1] Some of these have very subjective characteristics, such as the opinions and biases of the individuals or groups that develop the policies or are affected by them. Others lie half-way between objectivity and subjectivity, for example anecdotal information. This type of information may come from truly representative cross-sections of well-informed experts or it can be based on frequent and highly visible claims made by relatively few, prejudiced individuals. Since anecdotal information pertinent to important current policy issues has a tendency to be passed on purely verbally and is often eventually quoted as fact, its reliability is questionable. At the other extreme, there is factual information based on actual measurements or objective observations. Such information is most useful for policy formulation because if that process is to be effective it should be based on complete and accurate knowledge of the dimension of the problems, of the features of the enterprises likely to be affected by the policies, and of the dynamics of the system. Frequently, such factual information is of a quantitative nature and in the form of statistical representations. It is true that statistical information can be misused. However, if properly presented and of good quality (i.e., of meaningful statistical significance and produced with adequate quality control), such information can be exceedingly useful for policy development. It can describe the current status of the system for which policy is being made; it can provide, through trend analyses, insight into how the system came to its current status; and, as the basis for models, it can make possible descriptions of likely future outcomes of various scenarios representing different S & T policy options. Thus, this important policy formulation tool, if properly developed and used, can enhance decisions and increase the probability for effective policies. It must be remembered, however, that quantitative information represents only one necessary input to good decisionmaking; others, such as good judgment, are also required but are difficult to exercise without a reliable information base.

Clearly, the capability for use of factual information in policy development depends on the existence of data systems which measure various dimensions on a regular basis. If such data systems have been in existence for many years, analyses of trends are possible. This is important since comparisons frequently provide a better basis for evaluations than examinations of absolute magnitudes. Such comparisons can be between different institutions, organizations, or countries, and they can involve different entities within coherent groups, such as fields of science, industrial product groups, or countries in the same state of development. One exceedingly useful comparison involves the examination of the status of a system of various points of time, namely, trend information. For example, it is very difficult to determine whether sufficient research and development (R & D) is being performed in a country or institution. No means have yet been developed to prescribe an optimum absolute level for such R & D activity. However, comparisons with R & D in other comparable countries or institutions, appropriately normalized and combined with trend analyses, are frequently used to determine future directions of R & D activity levels. (See Figure 5.1.)

For the reasons just enumerated, there is probably very little disagreement among active S & T policymakers, policy analysts, or students of S & T policy that a good quantitative data base is a useful, if not essential, component of any science and technology policymaking apparatus. This essay examines various features of such data systems and will focus on those systems that are related to the operation of the S & T enterprise. These generally cover information dealing with input, output, and impact measures and are sometimes referred to as S & T resource information systems. Examples cited have been selected from the American S & T policy scene but are quite likely representative of similar issues in other countries.

Uses of S & T data systems

One of the primary purposes of extensive S & T resource statistics is the provision of a readily available, easily accessible data base for use by S & T policy analysts and decisionmakers in the course of their daily activities. Information is most frequently used in a reference mode. Such use can involve: verification of impressions, such as whether the age distribution of the technical work-force has changed significantly; checking of distributions, such as the relative role of various sectors of the economy in the performance of R & D or determination of absolute or relative magnitudes, such as how much money the government invested in academic research in a specific year. Used this way, statistical tables, charts or graphs serve policymakers functionally just as nuclear cross-section tables serve

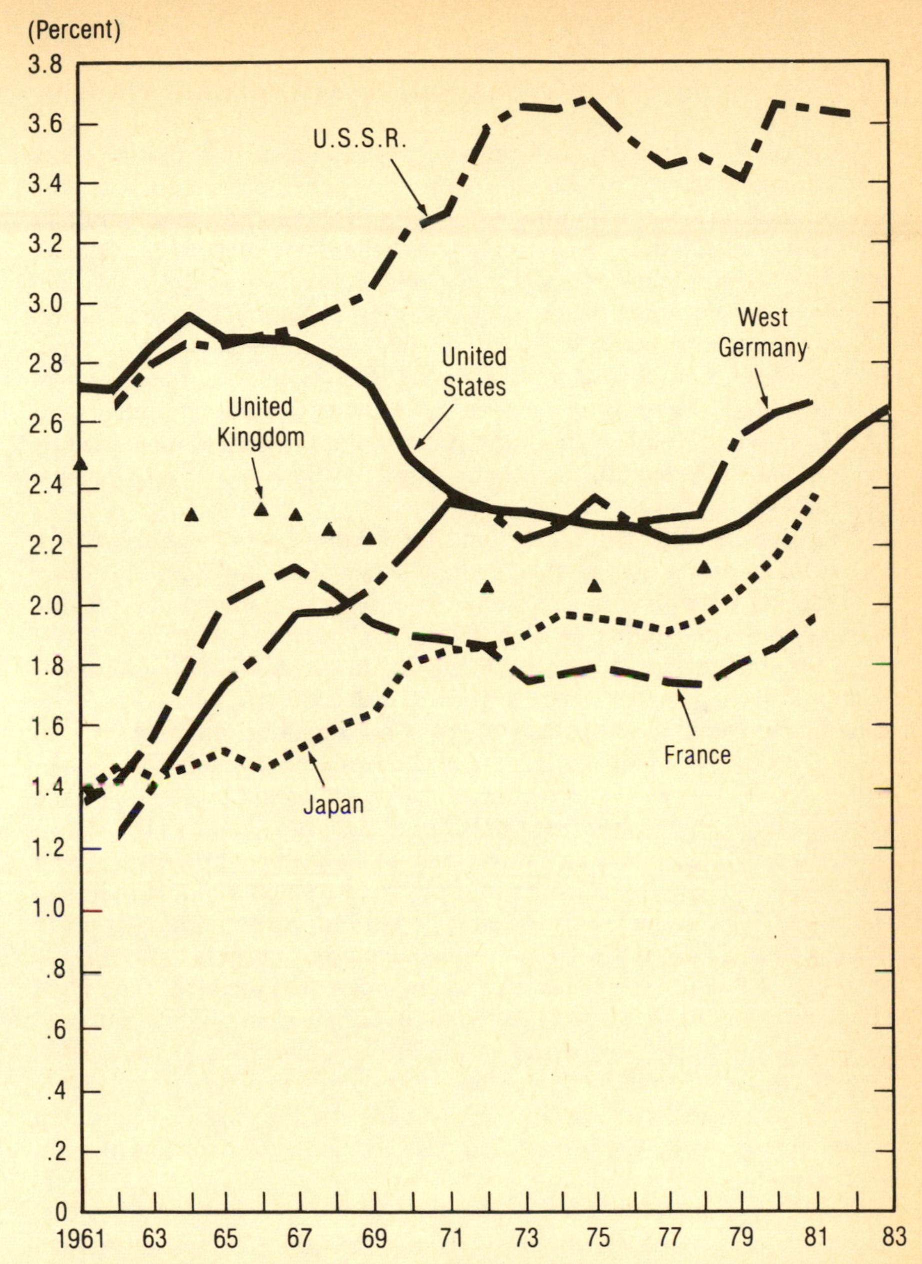

* Gross expenditures for performance of R & D including associated capital expenditures, except for the United States where total capital expenditure data are not available. Detailed information on capital expenditures for research and development are not available for the United States. Estimates for the period 1972–80 show that their inclusion would have an impact of less than one-tenth of one percent for each year.

Note: The latest data may be preliminary or estimated.

Source: Science Indicators—1982.

Figure 5.1 National expenditures for performance of R & D as a percentage of gross national product by country.

physicists, national income tables serve economists, or material characteristics tables serve engineers.

Data can also be used for different types of analytical activities. They can illuminate issues; for example, questions have been raised about the magnitude and status of equipment used in R & D and other scientific or technological activities.[2] Answers to these questions have led to national, regional or institutional policy changes related to future equipment acquisition. To develop such policies in a meaningful fashion requires data on magnitude, age, technological obsolescence, use and cost of various types of equipment[3] as well as related data on such parameters as magnitude of research budgets or of personnel groups using the equipment, or effectiveness of groups with different equipment resources measured by such quantifiable features as numbers and citations of research papers, patents, etc.

Data analyses can identify problems before they become highly visible, i.e. in their early stages of development when they may not yet be very evident. For example, constant monitoring of characteristics of academic science and engineering faculty (Figure 5.2) revealed that the ratio of recent-to-total doctorate faculty in the United States was beginning to decrease during the early part of the 1970 decade. This was repeatedly called to the attention of policymakers at both national and institutional levels,[4] produced a fuller awareness of the evolving problems, and eventually resulted in some corrective steps.[5] This example illustrates the importance of the maintenance of a broad data base prior to the occurrence of specific problems since absence of earlier benchmarks makes it impossible to detect deterioration or improvements once they are suspected. Trend analysis in general constitutes an important input to S & T policy. Thus, in the mid-seventies, analyses of trends in funding of academic R & D in various fields of science demonstrated that the rate of change in funding of the physical sciences and engineering had not kept pace with the funding increases of the life sciences. (See Figure 5.3.) A direct result of the examination of these trends was an overt Federal policy to accelerate funding increases in engineering and the physical sciences, a policy which has been maintained for a number of years by several administrations.[6]

Statistical information can be extremely useful to check out anecdotal assertions. For example, in the early eighties there were widespread reports that American institutions of higher education were encountering great difficulties in maintaining engineering faculties. Reports claimed that there were two reasons for this problem. The universities were allegedly not able to fill openings on their engineering faculties because of strong recruitment competition from industry caused by higher industrial salaries and better-equiped modern laboratories. These last two factors

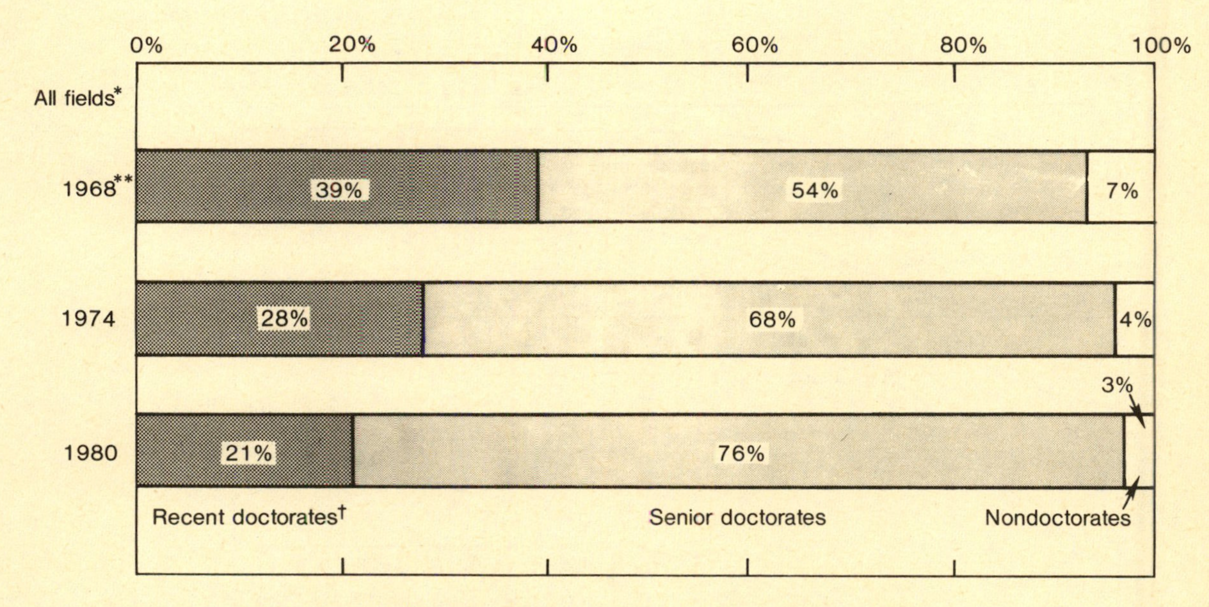

* Only the twelve fields are included that are common to the surveys for 1968, 1974 and 1980.
** Percentages for 1968 are based on unweighted data.
† Faculty who have held doctorates for seven years or less.
Source: National Science Foundation.

Figure 5.2 Trends in doctorate composition of full-time faculty in doctorate-level departments.

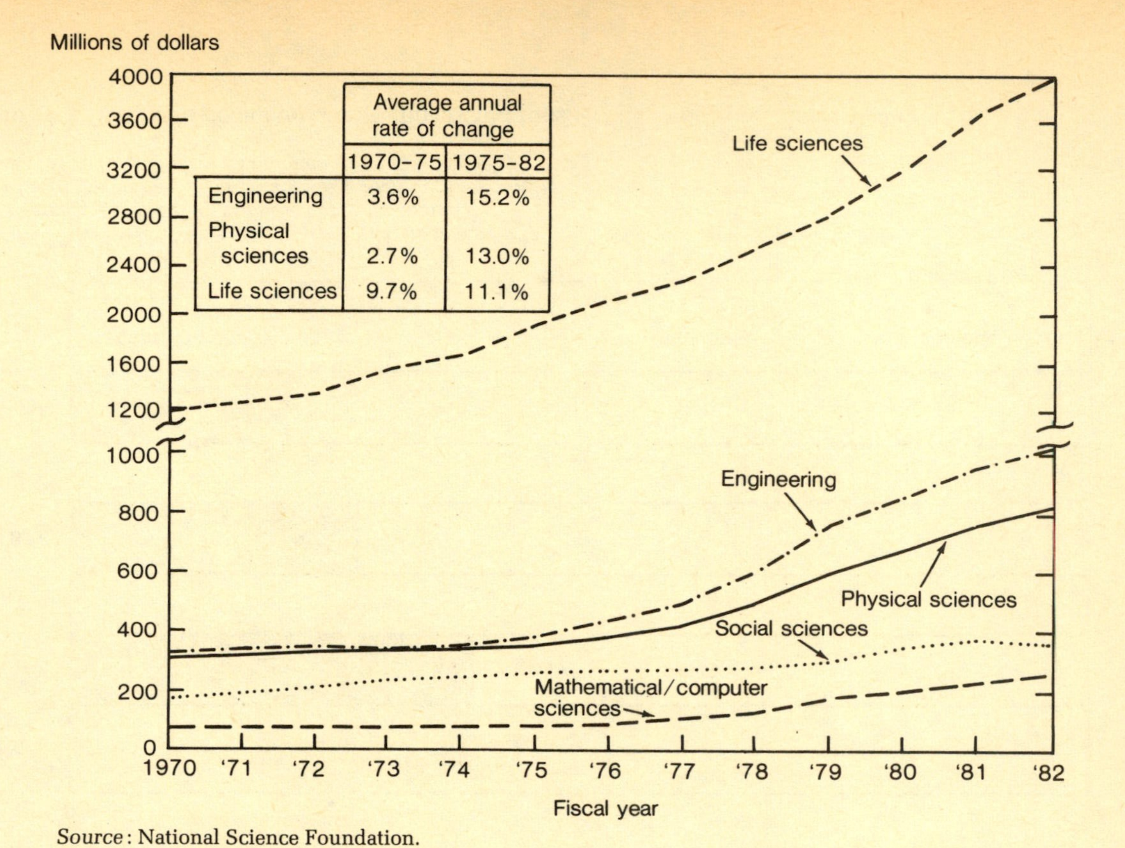

	Average annual rate of change	
	1970–75	1975–82
Engineering	3.6%	15.2%
Physical sciences	2.7%	13.0%
Life sciences	9.7%	11.1%

Source: National Science Foundation.

Figure 5.3 R & D expenditures at universities and colleges by S/E field.

were also alleged to cause engineering faculty members to leave academia in droves to accept industrial positions. A special survey of engineering departments[7] confirmed the first phenomenon: about 10 per cent of engineering faculty positions were unfilled at the time of the survey and 4 per cent had been open for more than a year. However, the same survey showed that the second alleged problem—the movement of faculty to industry—was relatively minimal, with only 3 per cent of engineering faculty having left voluntarily during the previous year for industrial employment.

A well-developed data system is essential for the development of models that try to describe how components of science and technology systems operate. Such components can be the technical personnel system, i.e. the supply and utilization of scientists and engineers; the funding system, e.g. the role of different sponsors of R & D or other innovative activities; or the operational aspects of a part of the system, e.g. the efficiency of different units judged by the relationship of inputs to outputs. It is possible to develop theoretical models of such systems which describe their operations in terms of interactions of various variables. However, only actual data make it possible to test the validity of the models and to assign coefficients that specify the strengths of linkages and variable interactions. One important characteristic of valid models is their ability to describe likely future behaviors of the system under various assumptions. These assumptions are generally estimates of likely possible changes in features that affect the system, including various proposed policy or action options. Such models, projections and scenario evaluations can be very useful to policymakers in a variety of ways. They can identify those factors that have the greatest effect in producing change; this narrows the field of possible policy options and prevents adoption of options that are not very likely to produce the desired results. Projections can also identify likely future problems as well as the reasons for the problems and thus lead to preventive actions. Similarly, and at least as important, projections can sometimes show that existing problems are not likely to persist for much longer and thus prevent the introduction of unnecessary corrective steps. This type of analysis is especially important when corrective actions have built-in lag times. For example, any action designed to increase or decrease the supply of technical personnel by encouraging or discouraging students to select curricula that lead to technical degrees produces results only years after the new policies have begun to be implemented. Projections using technical personnel models can illustrate whether the supply–demand imbalances which are to be corrected are likely to be still in existence at that future time. If not, the new policies could actually exacerbate the problems which they hope to ameliorate. Examples of such counter-productive actions can be noticed in technical personnel policies dealing with fields where student enrolments are very market-sensitive, such as engineering in the United States.[8] Since markets for engineers are quite dependent on economic cycles, past attempts to modify the number of new engineering graduates have frequently been sufficiently out of phase to worsen future supply-demand problems instead of solving them. As Figure 5.4 shows, labor market models have many components and require broad data bases. A recent forecast which evaluated the adequacy of the supply of science, engineering and technician personnel for various future defense and non-defense requirement scenarios,[9] provides a good illustration of the extensiveness of data required for human resources projections.

Before leaving the topic of projections, it is necessary to emphasize the importance of an awareness of their limitations. These can be due to the quality of the data that are used or the validity of the model itself. Furthermore, it must be recognized that projections are conditional forecasts that depend very heavily on assumptions that have been made. Such assumptions and descriptions of the model used should always be presented quite explicitly.[10]

Types of quantitative S & T resource-related information

The systematic collection of national S & T resource-related data began in the United States in the early part of the 1950 decade. The start of such activities can be directly attributed to the realization, arising out of World War II experiences, that science and technology were likely to be rapidly growing activities with ever-increasing impacts on national welfare.[11] The first efforts were designed to assess the overall nature and magnitude of the national S & T effort. It was found that only one type of feasible measure was available for such assessment, namely information on scientific and technical personnel (STP). Surprisingly, now, almost thirty-five years later, STP data still constitute the only comprehensive quantitative source of information that covers all S & T activities.[12] Shortly afterwards, data collection systems were organized to measure financial resources allocated and used for research and development. These R & D funding data provide not only information on the sources of funds for such activities but also serve as surrogate measures for activity levels, both total national and at the R & D-performing sector level. However, in order to be useful for activity measurement funding data have to be deflated. While some attempts have been made to develop deflators of R & D expenditures,[13] most have been relatively crude, using surrogate deflators for cost categories that are similar to those found in R & D financing. Consequently, applications of such deflators have to be handled with care, used only at fairly high levels of aggregation, and even then only with the

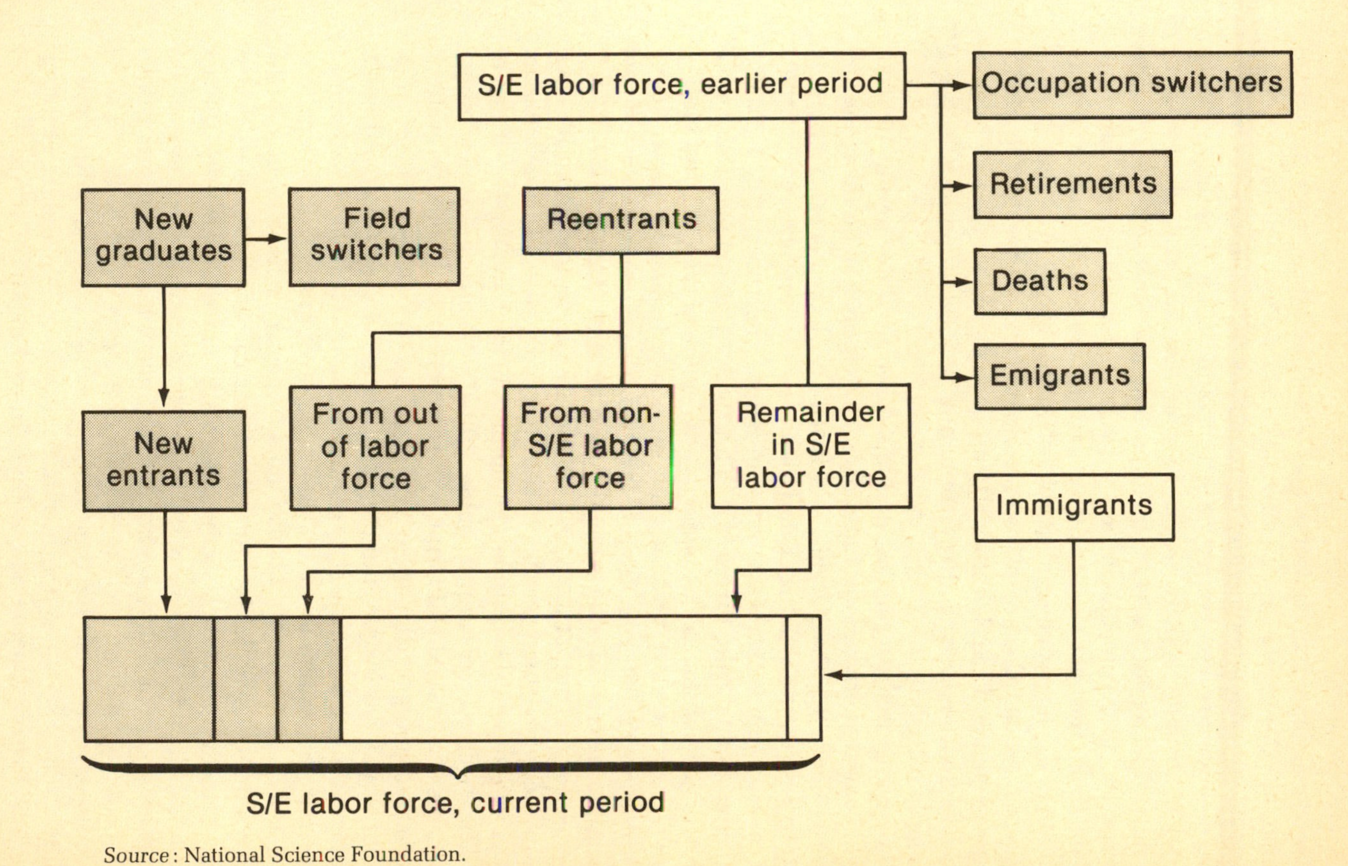

Source: National Science Foundation.

Figure 5.4 Science/engineering (S/E) labor market flows.

realization that the resulting constant currency funding levels represent only approximations.

The development of funding and personnel data systems spread throughout the world as science and technology policies received increased national attention and the need for such information became evident in country after country. Since determination of optimal or even adequate absolute activity levels is extremely difficult, comparisons among similar countries play an important role in many S & T policy discussions. Such comparisons are not very meaningful unless data from various countries are conceptually and definitionally similar or are adjusted to produce such similarities. International organizations, such as UNESCO and OECD, have played important roles in developing guidelines and standardized definitions[14] and have played very active roles in the collection, analyses and dissemination of R & D funding and S & T personnel data.

The quantitative information developed and used during the 1950–70 period consisted primarily of input data, i.e. information on the resources made available for the pursuit of science and technology. A common mistake is to assume that such inputs are also representative of outputs and to use input data as surrogate output measures; yet, this was frequently done during this period. The 1950–70 period was also one of continuous, substantial expansion of national S & T capabilities. However, by the early seventies the inevitable happened and in many countries growth patterns began to level off. This phenomenon produced increased interest in the measurement of the efficiencies of S & T systems—for this output measures are a necessity. In the search for such output measures there has been some conceptual confusion between output and impact. The former should only involve the direct outputs of S & T activities; for example, publications, citations and other bibliometric measures are direct outputs of research; patents and innovations represent outputs of the more complete S & T process. Use of such measures provides insight into the effectiveness of the system, although, except on a micro-basis, the state of the art is not yet sufficiently advanced to permit the examination of direct input–output relationships, i.e. productivity, at aggregate levels. Impact measures help to assess the effects outputs have. While output measures can generally be attributed rather directly to technical activities, this is not the case with impact measures, which try to measure the impacts of science and technology on the societies in which they are imbedded. Such impact indicators can be of an economic nature, such as balance of trade in technology-intensive products or royalty and fee transactions, or they can deal with various quality-of-life aspects, such as health, environment or standard of living. Even a relatively casual review of these impact measures will show that science and technology constitute only some of the factors that affect them. They do show the effects of S & T but they are one step further removed from direct S & T outputs and generally it is much more difficult to determine precise relationships between inputs and impacts.

The difficulties experienced with establishing direct relationships between inputs, outputs and impacts led in the decade of the 1970s to the development of a new technique for the use of quantitative information in S & T policy formulation, namely the development of the S & T indicator method.[15] The concept of indicators is subtle and differs significantly from that of plain statistical information. Science and technology indicators are quantitative measures of various facets of the S & T enterprise, which are frequently based on combinations of data and which often, taken one by one, constitute only partial measures of specific entities. Furthermore, many indicators depict what their name implies: they infer instead of measure. For example, R & D expenditures expressed in current currency terms provide inferences of activity levels. Many indicators have weaknesses in that their interpretation produces various degrees of ambiguity. The strength of the indicator method arises out of the use of numerous indicators, each frequently with a different weakness, to examine a particular input, output or impact characteristic. Taken together, such indicators can frequently lead to definitive conclusions about relative magnitudes, broad trends or other comparisons. A review of S & T indicators illustrates very rapidly that it is the exception, not the rule, for a single statistic or even a single statistical time series to be useful as an indicator of any significant element of the S & T scene.

S & T indicators are no panacea; for a variety of reasons, by themselves, they will not solve all or even most decision-making problems. The state of the art is still in its infancy and many more years of conceptual development, research and data generation are required. Also, many aspects of science and technology systems cannot as yet be quantified—possibly never. Finally, the systems under examination are complex and operate under the influence of many interactive factors. Frequently, there are no adequate methods to disentangle these causal features and to identify the effects of single factors. However, S & T indicators represent a very useful tool for policymakers when used within their limitations.

Types of data systems

In considering desirable characteristics of data systems for policy formulation, it is first necessary to distinguish between different types of systems. Clearly, those designed for institutional policy development are somewhat different and more detailed than those used for the broader regional, national or international scene. However, all have much in common.

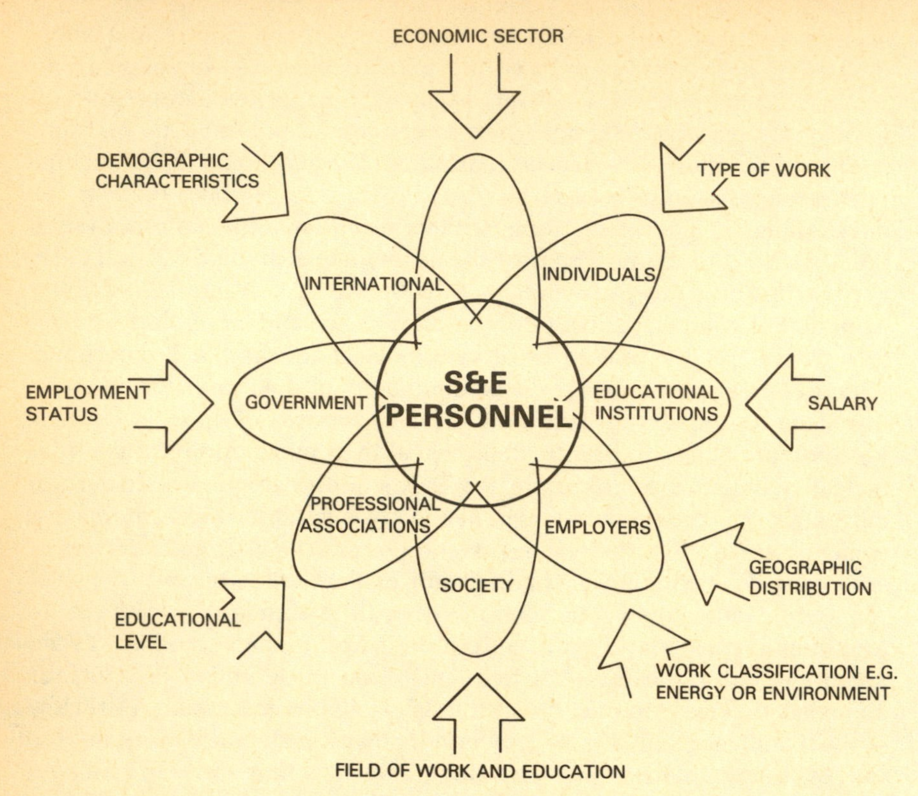

Figure 5.5 Characteristics of science and engineering personnel systems.

The core of any of the systems has to be a regularly updated set of data covering the principal parameters that describe the scope and operational characteristics of the system. In the case of personnel data this core should include information on: training and education, e.g. date, level and field of degree; current occupation; demographic characteristics, e.g. age and sex, employment, e.g., type of employer, work activity, current employment status; characteristics of the pipeline for future professionals, e.g. number of students in each field, their characteristics, and sources of support. (See Figure 5.5.) In the case of R & D funding systems, core information should cover: sources of funds; recipients of funds, i.e. the performers of R & D; character of R & D funded, e.g. basic research, applied research, development; field of research, e.g. chemistry or economics; product group of applied research and development, e.g. fabricated metal products or aircraft and parts; functional classificaiton of R & D, e.g. health-, defense- or agriculture-related. (See Figure 5.6.) In all of these cases, information on characteristics of related institutions must also be compiled so that various funding or personnel data can be analyzed in terms of their relationship to these characteristics. The type of institutional information to be assembled depends, of course, on the nature of the sectors in which the institutions are embedded. Examples are: type of industry; size; profits, sales and other economic characteristics for industrial institutions; or size, type of control (public or private), highest degree granted for post-secondary education institutions. Most of the data elements enumerated above are frequently required in most S & T policy discussions.

It is, of course, impossible, impractical and uneconomic to maintain regular data systems which contain all information which is likely to be required by policymakers. Frequency of anticipated need should be a major criterion in the selection of data elements for regular, periodic surveys. However, the nature of S & T policy issues does vary with time, new issues arise and old ones become obsolete in that they no longer command major attention. There are two ways to deal with this phenomenon. Periodic reviews, including user surveys, can identify obsolete and new data elements which are judged to be needed over extended periods of time. These can then be deleted from or added to regular surveys. On the other hand, data needs frequently arise out of issues which may be relatively temporary or, at least, not clearly of a persistent nature. In these cases the *ad hoc* survey is the solution. Such surveys can be tailored to the needs of very specific policy issues. If the issues continue to demand attention after an *ad hoc* survey has been carried out, then some data elements of the survey can be incorporated into the questionnaires of existing regular surveys or, if necessary, the *ad hoc* survey itself can be institutionalized as a regular survey. Clearly, the latter step should be taken only after very careful consideration, since it generally involves a significant addition to the workload of statistical units as well as to the burden of those who have to respond to questionnaires.

All the types of surveys discussed up to now are of the 'standard' type in that they require careful preparation, determination of feasibility, advance interactions with potential respondents, careful statistical design, development of uniform definitions of data elements which can be used by all respondents, form design, follow-up with respondents to assure adequate response rates, editing and, if necessary, follow-up with respondents to assure good data quality, extensive tabulations and analyses. Unfortunately, such a thorough and extensive process takes time and data from this type of survey are seldom available earlier than about twelve to fifteen months after the end of the time period for which data are collected. While this seems long to the uninitiated, it should be remembered that even institutions themselves generally do not have final information on their own records until several months after the end of the

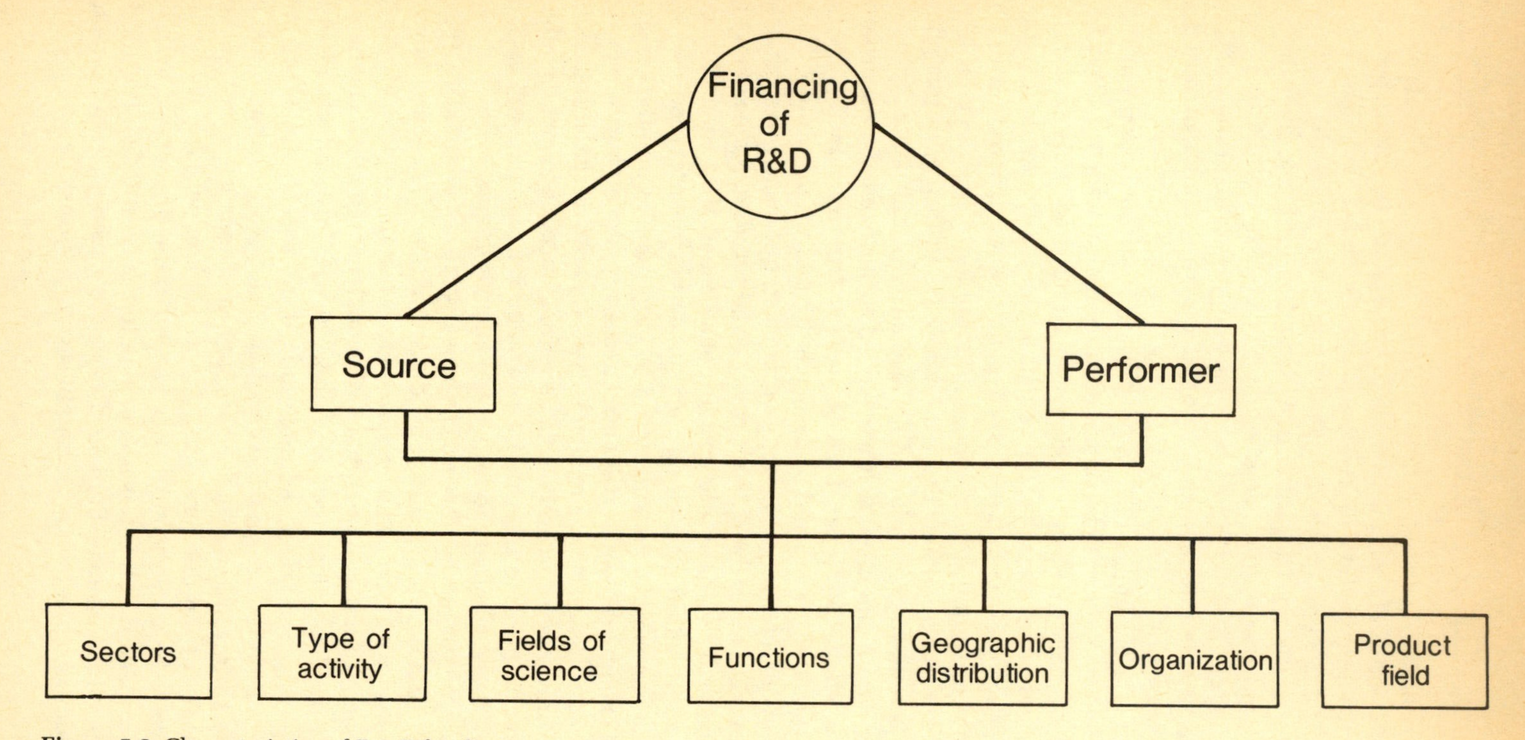

Figure 5.6 Characteristics of R & D funding.

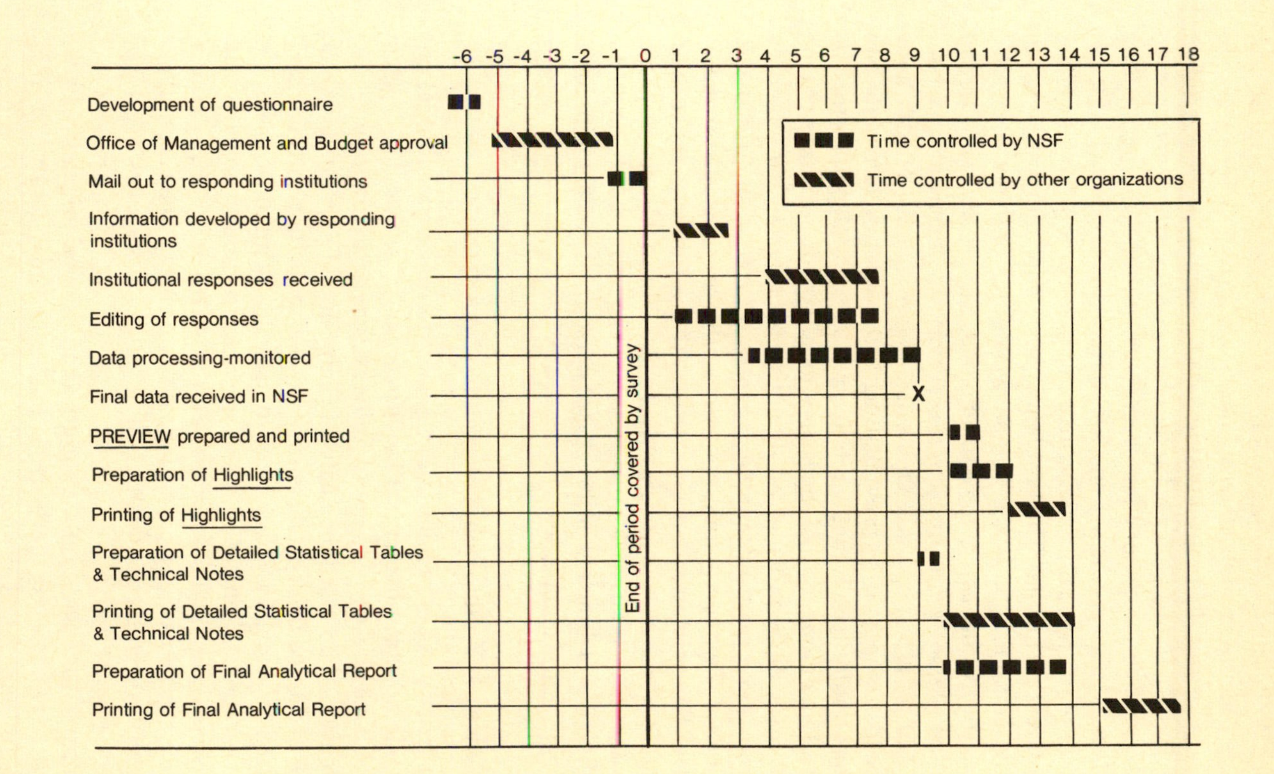

Source: National Science Foundation.

Figure 5.7 Average processing time for academic surveys of the National Science Foundation. (Time in months.)

time period under consideration. Figure 5.7 shows the various steps involved in a typical survey carried out by the US National Science Foundation. While some of the bureaucratic control steps may not be universal, most of the substantive steps are necessary anywhere for surveys of high quality. The time delays inherent in data collection can be a problem since many policy issues either arise suddenly or policymakers want to attack problems as soon as they focus on them without waiting for long periods before taking specific actions. Of course, if the required information is part of regularly maintained core data, there is no problem unless very current data are needed. The problem arises if new information has to be generated on an *ad hoc* basis. The answer to this problem is a rapid-response survey mechanism. Such a mechanism can involve collection of information by phone from a representative, though not necessarily statistically valid, sample of respondents. Or it can consist of panels of institutions which agree to respond, up to a previously agreed frequency, to a limited set of questions on information that is relatively easily available to them. The fact that in all likelihood most of the issues are also of direct interest to the respondent institutions provides an incentive for them to participate in rapid response surveys, results of which should, of course, be provided to each respondent. Two such rapid reponse panels are currently utilized by the National Science Foundation. One involves a thousand academic institutions,[16] all of which are not necessarily involved in each rapid-response survey; another is composed of individuals with specific related corporate responsibility (R & D directors or director of personnel) in 300 major high-technology industrial companies.

Characteristics of effective policy-orientated data systems

Within the framework of regular periodic and *ad hoc* data compilations and collections, what characteristics should an effective data system have? Many have already been enumerated in previous paragraphs.

The system must be based on a careful evaluation of user needs, which can only be identified from extensive interviews and a thoughtful explicit determination of who the primary users of the data are intended to be. The concept of primary clientele is used since most comprehensive data systems have and should have a wide spectrum of users.[17] Furthermore, if surveys are based on voluntary response, the needs of the respondents must be given careful consideration if high response rates are to be achieved and maintained.

The quality of the data should be high, though compromises between quality and timeliness will have to be made. In order to achieve high quality, definitions of data elements must be such that they can generally be uniformly applied by most respondents. This means that respondents either already have the data in a form which closely adheres to the definitions or that they should be prepared to start new data systems to generate such information. The latter does not happen very often unless the data needs of the surveyor corresponds to a new data need of the institution. More often, the second occurrence takes place first and the surveying organization can then take advantage of it. Response rates should be adequate to assure acceptably low levels of non-response bias and to minimize the need for excessive imputations. For many years surveys sponsored by the United States government required response rates of at least 60–70 percent. Quality control should be high. This can be achieved through careful pre-testing of survey concepts, definitions and questions to ascertain that respondents are likely to be able to provide meaningful data. While a necessary prerequisite, this alone is not sufficient. Editing of responses, which can be greatly facilitated by the use of computers, is necessary to identify fairly obviously questionable answers. These should then be checked out through personal contacts with the reporting individual. Finally, specific response analyses, involving interviews of respondents to determine the sources of data used for their response, can frequently identify weaknesses in questionnaires and thus in the collected data.

Responsiveness to information needs, both in terms of timeliness and comprehensiveness of data, is absolutely essential. However, this characteristic is one which can never be developed to a point where it will satisfy all clients. Furthermore, responsiveness is frequently constrained by: available resources (money and personnel), the ability or willingness of respondents to provide information, and the nature of the mechanisms available to collect the data. Operators of data systems must face the fact that many users would like to have data the moment they ask for it and that specialized user groups will want ever more highly disaggregated data. Users of data systems, on the other hand, should be helped to understand the inherent limitations of data collection processes.

This clientele–producer dichotomy points towards an important, but often neglected, operational characteristic of good statistical organizations, namely extensive interaction between the producers of data and their users. Such interactions almost always have to be initiated by the producers. They are essential for many reasons. They serve to identify the types of information required. Not only do these change as a function of time, but frequently policymakers have only vague concepts of the kind of information that they are likely to need. By discussing problems and issues, data-orientated information producers can often assist users in crystallizing their specific data needs. Such interactions can also help the data producer in anticipating specific data needs that are likely to arise in the future as policy discussions focus more specifically on issue definition and action alternatives. Thus, a good S & T statistical unit must interact with users either through specifically arranged interviews or through observations of S & T policy discussions. Such activities call for particular qualifications of personnel of S & T statistical units. A good many of the members of the unit should, either through their educational background or work experience, be able to understand the nature of S & T policy and the way the S & T enterprise operates. Thus, such groups must reflect not only good statistical and analytical expertise but also a firm understanding of the S & T system and its operation. Clearly this calls for a multi-disciplinary staff orientated towards interdisciplinary team efforts.

Impacts of S & T data systems

The impact of S & T resource data systems is difficult to ascertain. There are a number of reasons for this. The most prominent one is the fact that the data are frequently used in a reference mode and thus become only one of many inputs to policy decisions and programatic development. Thus, inquiries of users about specific actions that resulted from the use of data will most often generate a reply indicating frequent use of the information but inability to pinpoint specific policy actions that were directly influenced. User surveys or reviews by user groups, however, represent one of the best mechanisms to improve the effectiveness of data systems. Such reviews almost always produce suggestions on how data, analyses and presentation of information can be improved and thus they provide very effective feedback mechanisms.

The utility of data systems depends on several factors. Some, such as timeliness and comprehensiveness, have already been mentioned. Equally important are the modes of information dissemination. The best information loses its value if it is not disseminated rapidly, widely and in a format that makes it easy for the reader to grasp the major features of the data. It is false economy to spend considerable resources on data generation but then to under-support manuscript preparation and data dissemination. In selecting optimal modes of information display, the characteristics and work patterns of various user groups have to be considered. Since high-level policymakers frequently constitute the primary audience for S & T resource information, means have to be found to present data and their analyses in formats that have high information density. For this purpose reports and publications have to be concise and to the point, generally not exceeding a few pages. Unnecessary technical detail can be omitted: what counts are bottom-line findings. The use of telegraphic style and 'bullet'—type highlights should be encouraged. Furthermore, busy executives are frequently not number-orientated, nor do they have time to scan tables of numbers to determine what they mean.

Consequently, the use of charts and graphs is very effective since they generally convey almost instantaneously the integrated principal features of the data.

The needs of other clientele groups should, of course, not be neglected and can be served through different modes of communication. Extensive data systems offer rich bases for analyses. Since the types of analysis are generally numerous and since the timelessness of most policy-related studies affects their utiltiy, it is important that newly collected data be made rapidly available to as wide a group as users as possible. Such quick data releases can be most easily achieved through the publication of complete data tables supplemented by careful technical notes that describe the statistical and definitional features of the information. If computer utilization by analysts is a fairly common practice, user tapes, if necessary with appropriate confidentiality safeguards, present very effective means to generate broader analytical utilization of data.

Statistical groups should be encouraged to analyze their own and related other data. Such analytical activities are important for several reasons. The developers of the information are probably in the best position to know the strengths and weaknesses of the data and thus are more likely to use it in appropriate ways. Furthermore, the involvement in such analytical efforts will heighten sensitivity to specific data needs and should prove very effective in assuring greater utility of new data collections. Such analytical efforts, however, do take time. Thus, release of preliminary information to policymakers and dissemination of data tables and tapes should not be delayed by such analyses. Rather, they should be carried out in parallel and issued subsequently in separate reports. Since these documents should contain new insights arising out of the data, they should either contain executive summaries or their release should be accompanied by separately published summaries. Such practices will ensure that policymakers, who are unlikely to read lengthy analytical publications, will still be informed of significant new findings arising out of detailed analyses.

Summary

In summary, quantitative information is a key ingredient for effective S & T policy formulation. The development of such information should not be handled in an *ad hoc* fashion. Rather, a science and technology data base should be generated and nurtured by a centrally located unit which can assure appropriate continuity, uniformity of definitions, coordination and planning. Furthermore, data must be amassed over extensive periods of time to provide an adequate basis for trend information. In order to be of maximum utility S & T data systems should be orientated towards an indicator mode with equal emphasis on input, output, and impact indicators. Data should be available for referential use by S & T policymakers. However, the quantitative information should also be analyzed to verify anecdotal assertions, to identify problems in early stages of development, to illuminate existing policy issues, to develop behavioural models of the resource systems, and to project likely future situations. Finally, information generated should be widely disseminated in a timely and concise fashion and in a style which makes it easy for non-scientists to absorb the main features of the presented information. If these prescriptions are followed and if newly-generated data are timely and related to current policy issues, then a quantitative data base will prove of continued value and utility in any science and technology policy program.

Notes

1. For example, see Alvin M. Weinberg, *Reflections on Big Science*, Cambridge, Mass., MIT Press, 1967, Chap. 3; and Harvey Brooks, *The Government of Science*, Cambridge, Mass., MIT Press, 1968.
2. For example, US House of Representatives, Committee on Science and Technology, Subcommittee on Science, Research and Technology, *1983 NSF Authorization Hearings*, 23, 25 February and 4 March 1982 and *1984 NSF Authorization Hearings*, 23, 25 February, 1, 8 and 10 March 1983, Washington, DC, US Government Printing Office.
3. National Science Foundation, Academic Research Equipment in the Physical and Computer Sciences and Engineering, 1984.
4. National Science Foundation, *Young and Senior Science and Engineering Faculty*, 1974; NSF 75-302, 1975; National Science Foundation, *Decline in Recent Science and Engineering Doctorate Faculty Continues into 1978*, NSF 79-301, 1979; National Science Foundation, *Science and Engineering Faculty with Recent Doctorates Fell to One-Fifth of Total in 1980*, NSF 81-318, 1981.
5. National Science Foundation Program Announcement, *Presidential Young Investigator Awards*, NSF 83-35, 1983.
6. US Office of Management and Budget, *Special Analysis, Budget of the United States Government, 1982, Special Analysis K*, p. 312, US Office of Management and Budget—*Special Analyses, Budget of the United States Government, 1977, Special Analysis P*, p. 289.
7. National Science Foundation, *Engineering Colleges Report 10% of Faculty Positions Vacant in Fall 1980*, NSF 81-322, 1981.
8. Richard B. Freeman, *The Market for College Trained Manpower*, Cambridge, Mass., Harvard University Press, 1971, Chap. 4.
9. National Science Foundation, *Projected Responses of the Science, Engineering and Technician Labor Market to Defense and Non-defense Needs: 1982–1987*, NSF 84-304, 1984.
10. Ibid. And for example, see National Science Foundation, *Projection of Science and Engineering Doctorate Supply and Utilization, 1982 and 1987*, NSF 79-303, Chap. 5, 1979.

11. For example, Vannevar Bush, *Science—the Endless Frontier*, 1948, reprinted by the National Science Foundation, 1960.
12. Charles E. Falk and A. Fechter, *The Importance of Scientific and Technical Personnel Data and Data Collection Methods used in the United States*, Paper for the OECD Workshop on the measurement of Stocks of Scientific and Technical Personnel, October 1981.
13. For example, see OECD, *The Measurement of Scientific and Technical Activities, The Frascatti Manual*, Paris, 1980, Chap. 4.
14. UNESCO, *Recommendations Concerning International Standardization of Statistics on Science and Technology*, November 1978.
15. For example, see National Science Board, *Science Indicators—1982: An Analysis of the State of U.S. Science, Engineering and Technology*, NSB 1983-1, 1983.
16. American Council on Education, *Higher Education Panel, What is it? What does it do? How does it work?*, Washington, DC, 1983; also American Council on Education, 1983 Higher Education Panel Revision, Final Report to the National Science Foundation, Washington, DC, October 1983.
17. Charles E. Falk, 'Guidelines for Science and Technology Indicators Projects', *Science and Public Policy*, February 1984, pp. 37–9.

6 The new technological wave and the developing countries: problems and options

Amilcar O. Herrera

Introduction

This paper intends to explore the consequences of the impact of the new wave of technological innovations in Latin America, and the options open to the region. The frame of reference of our analysis will be the Kondratiev–Schumpeter theory of the 'long waves' of the world economy, a subject to which Christopher Freeman has made a considerable contribution in recent years.

For an adequate evaluation of the possible effects of the new long wave on Latin American countries, it is necessary to understand the present situation and its causes. In other words, we have to examine the impact of the previous 'Kondratiev cycle'—the one that started in the recession of the thirties, and culminated in the sixties—on the region. On the basis of this analysis, we will try to identify the main elements of a strategy for the future.

In an article published in 1935,[1] Kondratiev concluded that 'on the basis of the available data, the existence of long waves of cyclical character is very probable'. He added, 'In asserting the existence of long waves, and in denying they arise out of random causes, we are also of the opinion that the long waves arise out of causes which are inherent in the essence of the capitalist economy'. Kondratiev, although pointing out the association of technological innovations with the long cycles, did not establish a causal relationship between them. It was Schumpeter who, in 1939, proposed a theory for the cyclical behaviour of the capitalist economy.[2] In his view, the cause of the cyclical behaviour of the economy is technological innovation, which is promoted by entrepreneurs. The Kondratiev long cycles are the product of a series of articulated innovations, each of them constituting an 'industrial revolution'. A long cycle, in this view, would be a succession of 'technological transformations that affects the economic system'.

In a recent paper, Carlota Perez[3] has made an outstanding contribution to the understanding of the mechanisms and dynamics of the long cycles. She starts by pointing out that, in Schumpeter's theory, the whole process of 'creative destruction' generated by one, or by a group of innovations, develops within the economic system conceived as a self-regulated entity relatively independent of the social milieu. Society is affected by, and affects the economic process, but it is mainly an environment. In Perez's approach, society is seen as basically composed of two sub-systems: the techno-economic, and the socio-institutional, the first having a much faster rate of change than the second. The structural crisis produced by a cluster of technological innovation is not only a process of 'creative destruction' in the economic sphere, but also involves deep transformations in the socio-institutional system. As the rate of response of the two systems is different, the long cycles are the result of the resistance offered by the socio-institutional system to the transformations taking place in the techno-economic system.

For Perez, the long waves represent different 'modes of development', which are the response to the appearance of successive distinct 'technological styles'. The modes of development stretch from trough to trough in each cycle, but the technological styles evolve from the peak of one cycle to the peak of the next. This is due to the fact that a new technological style begins when the previous one approaches the limit of its possiblities. An important point in Perez's theory is that the final form that a mode of development will take does not depend solely—or almost solely—on the characteristics of the new technological style: 'The final form the structure will take, from the wide range of the possible, and the timespan within which the transformation is effected to permit a new expansionary phase will, however, *ultimately depend on the interest, actions, lucidity and relative strength of the social forces at play.*[4] (My italics.)

We are aware that the set of hypotheses so briefly summarized above is still under discussion, and that many gaps must be filled before a complete picture of the dynamics of the cyclical behaviour of the economy can be built. The questioning of present theories today ranges from the notion that cycles are an inherent characteristic of capitalist economy, to the validity of systematic correlations between economic cycles and waves of innovations.[5] Nevertheless, we accept this theory as the frame of reference of our analysis, for two reasons. The first is that it constitutes a conceptual construction that is internally consistent; it does not contradict the facts and, in our opinion, it has more explanatory power than any other proposed theory or hypothesis. The second, but more important, reason is that we are at present confronted by a recession of the world economy that has all the characteristics of the down-swing of a long cycle, associated with the emergence of a wave of technological innovations which is deeply affecting the structure of the whole economic system. So, whatever the difficulties we may have in applying the theory to past long waves, there is little doubt that it is the best instrument we have with which to understand the present one.

The new wave of innovations

As is well known, the first Kondratiev long cycle was based on the steam engine and the textile industry; the second, on the railroad and the metal-mechanical and steel industries; and the third, on the internal combustion and electrical engines, and on chemical industry. In each of these cycles, the whole profile of the productive system was transformed from energy and transportation to final consumption goods. The dominant characteristic of the 'new' wave—centred on microelectronics—is that its impact seems to be more important to the organization of production, the labour process, and the social division of labour, than to the general profile of the productive system. The Industrial Revolution, with the first great modern wave of technological innovations and the emergence of the proletariat, consolidated the capitalist economy, and changed Western society. The subsequent technological waves changed the whole profile of the productive system, but did not alter significantly the structure of capitalist society; this new wave, in our view, will affect the very basis of our society, as can be seen by considering briefly its central protagonist: the process of automation and robotization.

It is clear that in the advanced countries the basic cause of unemployment is the fact that every day we need less labour to produce the same amount of goods and services. This historical tendency that started to intensify almost from the beginning of the post-war period, but whose effects were to a great extent concealed by a high rate of economic growth, will be enormously accelerated by the progress of microelectronics.[6] Its main impact will be in the social division of labour. The suppression of most forms of physical or routine work will gradually eliminate the proletariat in the Marxist sense, because the role of wages will fundamentally change.

In all modern societies, access to goods and services is essentially conditioned by wages in the widest sense: the remuneration of personal work in any of its forms. In the future, this central role of wages will decrease—firstly, because one of the consequences of automation, by eliminating most jobs which do not require 'non-programmable' skills or creativity, will be to obliterate most significant forms of hierarchy in the labour process; secondly, because participation in the productive system will become a diminishing fraction of total human activity, and so will gradually disappear as the central determinant of distribution.

The transition to the new 'mode of production' will undoubtedly take a long time to be completed—of the order of two or three generations—but its first effects are already with us. The duration and characteristics of the cycle will basically depend on the response of the socio-institutional structure to the changes—or to the range of options of change—induced by the new technological style. In this cycle the effects of automatization and robotization will be, in our opinion, the most difficult to absorb by the socio-institutional structure, and consequently they constitute the main element of our analysis.

The impact of the new wave in the advanced countries

Virtually all forecasts of levels of unemployment for the advanced countries predict a continuous worsening of the present situation.[7] The institutional response that the advanced societies are giving to the growing problem of unemployment is based primarily on the payment of a 'salary' to the unemployed through social security services. The non-institutional response is a rapid growth of the service sector, and of the so-called 'informal sector' of the economy. Both types of response can only be transitory, and are mainly symptoms of the lack of an articulated strategy for confronting the problem. However, the growing recognition that the character of present unemployment confronts the advanced countries with a problem that cannot be solved without a complete questioning of the relationship between technology, employment and work, is leading to proposals that, although still very general, point in a fresh direction.

In a recent paper, Prof. W. Zegweld summarizes well the basic philosophy of this new approach:

> The problem is to organize the breaking down of barriers between traditional wage-earning employment and work in the widest sense of the term. Such work can provide income but also offers a social role, contact with others, an opportunity for creation or enterprise. It must not be proposed in a single, rigid setting identical for all but must be flexible enough to meet the wide variety of demands and respond to freely expressed choices. Instead of offering everyone a problematical full-time job, the aim is to allow everyone to find and choose a job in which working hours, level of pay and social security coverage are no longer pre-determined and closely linked but can be adapted, above an indispensable minimum, to the wishes of the individual. . . . This strategy . . . presupposes a close link between technological and social innovation, allowing greater flexibility in the organization of production . . . without neglecting an adequate level of protection for everyone.[8]

The problem is not whether or not traditional forms of work and employment will be abolished; that change is inherent in the socio-economic transformation induced by the new technologies and is, as such, irreversible. In the words of André Gorz: 'The choice is between the liberatory and socially controlled abolition of work (in the traditional sense of wage-earning employment) *and* its oppressive and antisocial abolition'.[9] We believe that the first option will prevail, partly because of its social and economic rationality, and partly because the oppressive imposition of new forms of work would be extremely difficult in a society where social control is largely based on discipline imposed by the traditional relationship between work and employment, precisely the relationship that will be radically changed.

The impact of the new wave in the Third World

To analyse the impact of the new wave of technological innovations on the countries of the Third World, and the options available to them, we will consider the Latin American case. The first reason for this selection is that the diversity of the Third World makes any global treatment too general. The second is that new technologies are entering faster and more massively into Latin America than into any other Third World region. What we can learn in this exercise can provide a frame of reference—or at least a general guideline.

It is not necessary to describe in detail the socio-economic development of Latin America—and most of the Third World—in recent decades. Its development was based upon the evolution of the developed economies, particularly in Western Europe, in the post-1945 period. The success of the Marshall Plan and the rapid acceleration of technological progress were associated in those countries with a period of prosperity without precedent in the history of capitalism.

Two elements—the influx of capital and the introduction of new technologies—were adopted by the dominant classes of Latin America as pillars upon which to sustain economic and social development. Beside the intrinsic advantage of those technologies, and the pressure of the advanced countries—basically through expanding multinationals—to disseminate those technologies, this strategy of development offered two important advantages. The first was its simplicity. It involved the mechanical translation of a conception originated in advanced countries and accepted on the basis of its demonstration effect. Secondly, it seemed to ensure economic growth—its association with social progress was taken for granted—without substantial changes in prevailing social and economic structures.

The results of this strategy are well known, and a few indicators are enough to describe, in general terms, the present situation. The GNP per capita of the region was 10 per cent lower in 1983 than in 1980.[10] The rate of inflation in the most industrialized countries of the region—Argentina, Brazil, Mexico, Venezuela, Chile—has reached values without precedent in the past, and the external debt of the region amounts to about $340 billion.

Those figures only indicate the overall situation, and do not reflect the most important results of the strategy of development. During most of the period we are considering, the rate of economic growth of the countries of the region was high: between 1965 and 1981 the GNP of the region quadrupled, while the GNP per capita doubled. However, the benefits of that sustained growth reached only a minority of the population, because the pattern of industrialization was mostly directed to the requirements of a bourgeoisie and a middle class with the same pattern of consumption as their equivalents in the advanced countries. The rest of the population was, at the end of the period, in a situation not much better than in the past, and sometimes even worse. A partial exception to this overall picture lies in the countries of the 'Southern cone' of the region—Argentina, Chile, Uruguay—where a better distribution of income allowed a greater proportion of the population to benefit from economic growth.

Latin America now confronts a situation which leaves little room for the superficial optimism prevalent at the beginning of the post-war period. The previous strategy of development based on the massive influx of capital and technology is no longer viable. Beside external debt, which makes it very difficult for the region to incorporate more foreign capital, the world recession is a fundamental factor in restricting the transference of capital from the centre to the periphery.

Summing up, it is clear that the wave of innovation associated with the previous long cycle failed to generate more and better-distributed wealth in Latin America as it did in the advanced countries. As a consequence, while the developed countries are entering a 'post-industrial' era, the countries of Latin America—as well as most of the Third World—are receiving the impact of the new wave without having received the benefits of the previous one, or of the Industrial Revolution more generally.

The reasons for this failure have been amply discussed in Latin America in terms of dependency theory, which places the main cause of the persistence of underdevelopment in the Third World countries on their mode of insertion into the international economic structure. A variation on this general outlook—with many advocates in the advanced countries—recognizes the role of international and social settings, but tends to stress the relative backwardness of developing countries in science and technology. According to this interpretation, an adequate capability in science and technology is an essential precondition to benefit from the process of change induced by waves of innovations. The absence of that precondition would explain, to a great extent, the failure of the Third World in the so-called process of modernization.

On balance, we believe that the available evidence supports dependency theory. However, in the context of the present analysis we want to focus on the direct agent or mechanism responsible for the failure to benefit from the previous cycle. A reasonable understanding of that mechanism is one of the essential preconditions to a strategy for the future.

In our view, the immediate cause of the poor performance of Latin American economies in the third Kondratiev long wave was their failure to adapt the socio-institutional sub-system to the changes of the techno-economic sub-system induced by the wave of innovations. This is suggested by the key socio-institutional transformations associated with the last long wave: the redistribution of income, the strengthening of labour unions and their institutional acceptance, and the internationalization of the economy.[11]

The redistribution of income in the developed countries as a consequence of the mass production technological style had two main effects: to enormously enlarge internal markets, and to change the pattern of product demand. The market which had previously been divided between luxury and staple goods evolved to meet the demand of the middle income sector which, including the middle class and considerable proportion of the workers, comprised the majority of the population. In Latin America there was no significant redistribution of income; in most countries, on the contrary, there was a continuous concentration of income in the upper classes. The most important consequence from the point of view of the productive structure was that the demand for non-staple goods came only from those minorities with an income equivalent to the upper and middle classes of the advanced countries. The result was that the pattern of production was not determined by the demand of the majority of the population as was the case in the advanced countries, but by the demand of that privileged minority. Thus, the imitative style of industrialization, with the concomitant massive and apparently indiscriminate transference of technology, was not a consequence of technological backwardness—even with the same basic technological elements, the composition of the final goods 'package' could have been different—but rather a response to the pattern of product demand.

Secondly, at the beginning of the period, we saw the industrialization of Latin American countries occurring with a small and poorly organized labour-force. Structural unemployment, aggravated in many countries of the region by the existence of a large, poor peasantry which migrated to the cities, put most of the industrial work force on the defensive. On the other hand, the chronic instability of the governments of the region made them suspicious of any potentially contestatory movement, and they used every possible means to repress or control trade union activities. Thus, despite a long history of sporadic struggles, the trade unions of Latin America—with the exception of Argentina, and Chile during the Allende Government—never had the social and political power of their counterparts in the advanced countries.

Finally, the internationalization of the economy with the rapid expansion of the multinational enterprises, and with the inter-country trade and investment regulated by international agreements, generated a new situation in world economic relations. The insertion of a country in the world system did not depend any more solely, or almost solely, on the market forces—controlled fundamentally by private enterprise—but also on the bargaining capacity of the nation states. Most of the Latin American states were, and are, intrinsically weak. Besides, the ruling classes whose vested interests were articulated with foreign interests have lacked the political will to fight for a more equitable world order, the only element that could have counterbalanced, at least partially, the superior political and economic power of the advanced countries. The natural result of that unequal struggle has been dependency, or 'neocolonialism'.

Latin American countries do not seem now to be in a better position than they were in the past to incorporate the new wave of innovations. On the contrary, their enormous external debt, with its paralyzing effects on the economy, the advancing world recession, and the social impact of automation and robotization, seems to make the incorporation of the new wave even more difficult.

Taking only the process of robotization and automation, a contrast between the advanced and the 'peripheral' countries is particularly revealing. The former, with low rates of demographic growth, high capacity for capital accumulation, and no structural unemployment until recently, cannot cope with the problem of unemployment which will probably reach the level of 20 per cent at the end of this decade. It is obvious that in Third World countries, with high rates of demographic growth, low capacity of capital accumulation and chronic structural unemployment, the employment situation will deteriorate much faster than in the advanced countries. This is particularly true in Latin America, where the large and middle-size countries are fairly industrialized.

The studies made in the advanced countries—the most important ones being Interfutures (OECD), the Presidential Report on the Year 2000 (USA) and the Brandt Report—seem to confirm a pessimistic view of the future of the Third World. According to those forecasts, the gap between the two worlds will be the same or greater than now in relative terms, or will diminish only marginally in the more advanced developing countries. In absolute terms, the situation in a great part of the Third World will probably worsen.

The question is, to what extent are these forecasts reliable? If we accept their implicit premises, they are, and those premises are essentially two: firstly, that there will not be radical changes in the present social and international structure, although they admit some adjustments, and possible changes in the pattern of distribution of power among the advanced countries. Secondly, that the Third World countries will be unable to take autonomous decisions that can change their mode of insertion in to the world economic order. The only exception to this assumption is one of the OECD scenarios which, based on the Latin American World Model,[12] supposes that the Third World countries will break—at least partially—their economic links with the big powers, and will implement a policy of self-reliance based on South–South co-operation.

In our opinion, these models of the future are too static and do not reflect the depth of the crisis, and the real scope of options opened by it. There are alternatives which could allow the region to benefit from the new wave of innovations.

A commonly-used classification of scenarios distinguishes *tendential* scenarios, describing a possible future assuming the persistence of recent main tendencies, from *normative* scenarios, which propose a set of possible and desirable objectives to guide actions. In our view this classification, although analytically useful, is somewhat misleading. To assume the persistence of present trends implies a decision to ignore the tendencies and potentiality of change inherent in any social system, and that decision is essentially 'normative'. The difference between the two groups of scenarios is really which type of 'norm' they use; in the tendential group, the norm is given by present trends, while in the other group, that role corresponds to a selected desirable and viable future.[13] In our treatment of the Latin American scenarios we will use the normative approach in *sensu stricto.*

Socio-institutional changes

We will begin with food production, a central element of any Third World scenario, whose main determinant is population. The average rate of demographic growth in Latin America is about 2 per cent, which implies that the region should more than double its food production in the next thirty years, taking into account that a considerable part of the population is now underfed. Such expansion of production—based mainly on increase in productivity, rather than on the incorporation of new land—should be guided by two main objectives: to provide food for the whole population, and to increase the income of the rural worker, whose cheap labour has until now been basic to the growth of the modern sector. The second objective is to contribute to the generation of the economic surplus required for the expansion of the industrial and service sectors. The transformation of the rural areas means a radical modification of the land tenure system, and the careful selection of the technologies and methods of production to be introduced. Although the combination of technologies and the organization of production should try to minimize migration to the cities—basically through a sustained improvement in the well-being of the peasants—the net result will be a decrease in the manpower required for rural activities, with the ensuing increase in the labour-force potentially available for the other sectors of the economy.

The key to any desirable scenario lies in ensuring that the benefits of the process induced by the wave of innovations reach the whole population. It can be estimated that more than 40 per cent of the Latin American population is outside, or almost outside, the non-staple goods market, and a considerable part of it even below an adequate level or provision of staple goods. The effective incorporation of the whole population into the market of goods and services, through a redistribution of income, represents a formidable social, economic and technological challenge. But, beside its intrinsic desirability, it is also the central component of a strategy that could allow the region to successfully confront the impact of the new innovations.

The main effect of redistributing income would be a change in the pattern of product demand, including services. The *average* income, which conditions the market, would be less than that in the advanced countries, and consequently the pattern of consumption of non-staple goods should be different. It would be a process similar to the one that earlier changed the pattern of goods demand in the advanced countries.

The second result of the process, by expanding the productive system, would increase employment by counterbalancing—at least in a transitional period—the effect of automatization and robotization. As in the case of agriculture, the selection and introduction of new technologies should be carefully controlled through a strategy adapted to the specific conditions of the region.

These socio-economic changes are very few, but should be present in any desirable society. Most important, they would allow the region constructively to confront the new wave of innovations. The employment impact of automation—socially, most difficult to absorb—would be delayed by the incorporation of a greater mass of population into the market. This would give Third World countries an important advantage over advanced countries. The fact that the former would have to change their patterns of consumption would give them the opportunity to adapt material consumption to a new concept of well-being, more in accordance with the resources and values of post-industrial society, where free time devoted to creative activities, and the preservation of an equilibrium with the physical environment, will be the principal characteristics. When the full impact of automation finally reaches Third World countries, this new conception of development, and the experience of the developed countries, could help them to absorb change with a minimum of social hardship.

This revision of the concept of development will necessarily also occur in the developed countries. We hope that this convergence will lead to a new world order based on solidarity and co-operation. The road will, however, be a long and difficult one.

The scientific and technological challenge

The proposed strategy represents a great scientific and technological challenge to the Latin American countries, and the weakness of their R & D systems is often considered the main obstacle they have to confront. However, the relative backwardness of the scientific and technological capacity of the region is not a cause, but rather a *result* of socio-economic development.

Our position, developed in detail elsewhere,[14] is that all countries have a science policy—explicit or implicit—which is determined by the character of its national project—i.e. the set of objectives (or model of the country) *to which the social classes which have direct or indirect economic and political control aspire.* History shows that when a national project demands an autonomous policy, scientific and technological backwardness is never an insurmountable obstacle, as is strongly suggested by the experiences of Japan, China and the Soviet Union.

On the other hand, new innovations offer developing countries the opportunity to enter those technological fields in their early stages of development. As it takes some time to transform innovations into transferable technological packages, this would allow, at least to the most advanced Third World countries, the development of technological versions more adapted to their own conditions and needs. This possibility is well illustrated by the success of Brazil in the arms industry.[15] Through an intelligent policy of exploring the specific needs of the developing countries, Brazil has become one of the main exporters in one of the most sophisticated and competitive technological fields of the world. If applied to civilian needs, there is no reason why that strategy could not be as successful as it is now—unfortunately—in the arms race.

In a recent paper, Giovanni Dosi develops the concepts of the 'technological paradigm'—a concept parallel to Kuhn's scientific paradigm—and of 'technological trajectory', or the way the paradigm evolves.[16] A technological paradigm is not a closed system whose evolution is unequivocally determined, but consists of a core of knowledge and basic technological elements which offer a variety of possible trajectories, whose direction is to a great extent determined by the social environment. The main objective of R & D systems should be the exploration of possible technological trajectories, in order to concentrate on those best adapted to the conditions of the region. This task, given the relative weakness of private enterprise and of the R & D systems of the region, will require strong support from governments—not only in terms of financial assistance, but through well-articulated policies with clearly defined priorities.

Towards the future

The proposed strategy for the future implies the introduction of radical changes in the socio-institutional structures of the Latin American countries. The obvious question is: which are the social forces able, and willing, to implement those changes?

One of the main results of the previous long cycle was the emergence of a middle class with some characteristics without precedent in the history of capitalism. That middle class is, numerically, in developed countries and in some Third World countries, as important or more important than the proletariat. It is also the repository of practically all the scientific, technological and administrative knowledge required by society. Thirdly, most of its members are wage-earners; in other words, they have a similar position in the social structure—although with better salaries—to the proletariat.

The impact of new innovations on that middle class will be as dramatic as it will be for industrial workers. Automation is entering into services which have always been basic fields of employment for the middle class, as fast, or faster, than in industry. Automation is already replacing the routine component of professional work, and is resulting in a growing army of unemployed university graduates. Besides, the advances of telematics, which poses the possibility of managing and controlling production units of the multinationals located in the Third World countries from the centre, will deprive the middle class of one of its main sources of high-level professional positions.

It seems obvious that there are common interests between that middle class and a labour force which is starting to understand—above all, in its most advanced sectors—that its main problem is how to confront a process of transformation that will radically change the traditional relationship between employment, work and technology. Unless an adequate social strategy can be implemented with the support of other social forces, the process could be as traumatic for the proletariat as was the introduction of mechanization in the first stages of the industrial revolution. Whether and how that class alliance or agreement can be articulated is difficult to forecast.[17]

Such a combination of social actors can bring the social consensus necessary for strengthening the nation state: one of the essential conditions for the success of Latin American countries in negotiating an equitable insertion into the emerging international order. It can also open the road to a democratic and participatory society which can mobilize the creative potential of the whole social body. In Christopher Freeman's words,

> Just as an individual combines intellect, imagination, reason, feeling and intuition, so must a healthy society blend the fragmental compartments and artificial divisions in our knowledge system and our professions. Only in this way can alienated technology become human technology. Innovation is far too important to be left to scientists and technologists. It is also far too important to be left to economists or social scientists.[18]

Notes

1. N. D. Kondratiev, 'The Long Waves in Economic Life'; *Review of Economic Statistics*, **18** (1935), pp. 105–15.
2. J. S. Schumpeter, *Business Cycles: A Theoretical, Historical and Statistical Analysis of the Capitalistic Process*, New York, McGraw Hill, 1939.
3. C. Perez, 'Structural Change and Assimilation of New Technologies in the Economic and Social Systems', *Futures*, **15** (1983), pp. 357–75.
4. Ibid., p. 360.
5. A general discussion on the subject of technological innovations and long cycles can be found in *Futures*, **3**, No. 4 (August 1981), with articles by Jos Delbeke, 'Recent Long Wave Theories: a Critical Survey'; J. Tinbergen, 'Kondratiev Cycles and So-Called Long Waves: The Early Research'; J. J. Van Duijn, 'Fluctuations in Innovations Over Time'; G. Mench, C. Coutinho and K. Kaasch, 'Changing Values and the Propensity to Innovate"; A. Kleinknecht, 'Observations on the Schumpeterian Swarming of Innovations'; J. Clark, Christopher Freeman and L. Soete, 'Long Waves Inventions and Innovations'; J. W. Forrester, 'Innovation and Economic Change'; E. Mandel, 'Explaining Long Waves of Capitalistic Development'. See also Christopher Freeman, 'Long Waves and Technical Innovations', paper presented to the International Symposium on Perspectives of Science and Technology Policy, Guanajuato, Mexico, February 1984.
6. For a good analysis of present trends see R. Rothwell and Walter Zegweld, *Technological Change and Employment*, London, Frances Pinter, 1979.
7. In a conference at the International Symposium on Perspectives of Science and Technology Policies (Guanajuato, Mexico, February 1984), Dr Ricardo Petrella reported in a recent study by the FAST Programme (Forecasting and Assessment in Science and Technology) of the Commission of the EEC which suggests that the 12 million unemployed today will grow to 20 million by the end of this decade.
8. W. Zegweld, 'Technology, Employment and Work', paper presented to the International Symposium on Perspectives of Science and Technology Policy Guanajuato, Mexico, February 1984.
9. A. Gorz, *Adieu au prolétariat*, Paris, Éditions Galilée, 1981, p. 17.

10. *CEPAL, Notas sobre la Economīa*, Nos. 387–8 (December 1983); A. Furtado, '1° Relatório de Andamento—Dinâmica Sócio-Económica de América Latina', Projeto Prospectiva Tecnológica para a América Latina, São Paulo, Brazil; CEBRAP, 1984, mimeo., pp. 21–5.
11. C. Perez, op. cit., note 3, pp. 368–71.
12. A. O. Herrera, *et al., Catastrophe or New Society?*, Ottawa, International Development Research Centre, 1977.
13. For general reviews of the subject of forecasting and scenarios, see L. Miles, 'World Views and Scenarios'; I. Miles, S. Cole and J. Gershuny, 'Images of the Future'; S. Cole, 'The Global Futures Debate, 1965–1976'; and S. Cole and I. Miles 'Assumptions and Methods: Population Economic Development, Modelling and Technical Change' in Christopher Freeman and M. Jahoda (eds), *World Futures: The Great Debate*, London, Martin Robertson, 1978.
14. A. O. Herrera, 'Social Determinants of Science Policy' in Charles Cooper (ed.), *Science, Technology and Development*, London, Frank Cass, 1973.
15. R. Dagnino, 'Technological Prospective for Latin America, Towards a Methodology of Analysis', paper presented to the International Symposium on Perspectives of Science and Technology Policy, Guanajuato, Mexico, February 1984; and 'A Indústria de Armamentos: O Estado e a Tecnologia', *Revista Brasileira de Tecnologia*, **14** (1983), pp. 5–17.
16. G. Dosi, 'Technological Paradigms and Technological Trajectories', *Research Policy*, **11** (1982), pp. 147–62.
17. This subject has been treated by, among others, A. Tourain, *Production de la Societé*, Paris, Seuil, 1977: Anthony Giddens, *A Estrutura de Classes nas Sociedades Avançadas*, Rio de Janeiro: Zahar, 1978; A. Gorz (ed.), *The Division of Labor*, New York, Humanities Press, 1976.
18. Christopher Freeman, *The Economics of Industrial Innovation*, London, Penguin, 1974, p. 309.

PART III

Technology and Employment

7 Strategic questions in social research: the case of unemployment*

Marie Jahoda

For the second time in this century mass unemployment has become a major social issue in the Western world and, just as during the Great Depression of the thirties, a considerable number of social scientists are currently engaged in studying this unwelcome feature of contemporary national economies. Several disciplines are involved; economics and social psychology have made most of the relevant contributions. That there are controversies in both fields, perhaps particularly in economics, is not surprising given a topic that bears so directly on fundamental differences in values, affects millions of people and is inevitably relevant for policies, however 'pure' the intentions of a theoretician. I do not intend here to enter into these controversies: in any case, I am not competent to do so with regard to economics. The purpose of this article is rather to examine the types of questions social psychologists have asked in tackling empirically the study of unemployment and the link, if any, between their work and that of other disciplines.

The number of relevant studies is by now so large that comprehensive coverage, even if it were possible, would be repetitious and hardly conducive to an overall assessment of the state of the field. Almost all the examples in the following pages are deliberately chosen from the very recent past on the assumption that investigators have taken earlier studies into account. Reviews of research results in a prolific field are inevitably selective.[1] An excellent survey by Kelvin and Jarrett,[2] for example, limits itself to four major questions:

- How does his unemployment affect the way in which the unemployed individual sees himself?
- How does his unemployment affect the way in which he sees others?
- How does his unemployment affect the way in which he perceives himself to be seen by others?
- How do others actually see the unemployed?

These are important questions but there are others. Even though Kelvin

* I am grateful to Lotte Bailyn, David Fryer and Keith Pavitt for helpful comments on an earlier version of this paper.

and Jarrett touch upon some of them in their illuminating commentary on the research literature, they remain implicit and unformalized.

In such reviews and in the studies themselves, social scientists go about their business of understanding the social world in a large variety of ways. Guided by various theories or simply by presuppositions and assumptions, they ask different questions, use different concepts and collect their data by different methods. Such differences are unavoidable and in the end salutary because they illuminate a topic from many perspectives. By the same token, the right of investigators to ask questions that seem important to them impedes the examination and perhaps the advancement of the entire field. For that purpose what is required is a formulation on a higher level of abstraction of the types of strategic questions that social scientists ask about a phenomenon, so as to acquire a yardstick by which limitations, if any, can be recognized. Accordingly, I shall first present two efforts to identify the universe of strategic questions that have so far been asked in the social sciences, then I shall use this to identify dominant trends and gaps in the collective effort, and finally discuss the needs for linking social psychological research to that of other disciplines as well as the difficulties that confront such efforts.

Formal strategies in social research

Several analyses of the logic of procedure in the social sciences are available; I have chosen two because of their compatibility. The earlier one is Robert Merton's paradigm for functional analysis;[3] more than thirty years later Gary Runciman produced the second, termed 'the methodology of social theory'.[4] The two schemes show considerable though unacknowledged overlap in several features, notwithstanding differences in terminology. Merton describes his scheme as a codification of procedures; Runciman describes his as a guide to practice—a clear identity of purpose. Both address themselves to empirical researchers who are aware of the 'triple alliance between theory, method and data',[5] both develop and justify their schemes through wide-ranging references to the research literature that exemplify various items of each scheme by good or bad practice, as the case may be; finally, the formal items in research procedure of the two schemes overlap—as they must unless both are mere figments of the imagination which, of course, they are not.[6]

There are other than terminological differences between these two codifications of formal questions, not least their respective degrees of differentiation. Since Runciman's is simpler in basic concepts (though not in exposition), it is convenient to begin with his 'guide to practice'. The crux of his scheme is an analysis of the concept of 'understanding'. He distinguishes three ways in which a social phenomenon can be said to be understood: first, in the sense of correct and adequate *reportage* of what has been observed. The major issues here are the choice of what to report and of how to classify it conceptually. Second, in the sense of what caused it or how it came about, i.e. *explanation*. Third, in the sense of knowing what it is like to be a participant in the situation under study, what he or she thinks, feels, says or does, termed somewhat idiosyncratically *description.* It is the inclusion of these last subjective factors that, according to Runciman, distinguishes the logic of procedure in the social sciences from that in the natural sciences—not a particular difficulty in arriving at explanations, as is often wrongly assumed. The criteria for good research under these three categories are accuracy for reportage, validity for explanation, and authenticity for description of experiences.

To this threefold guide to understanding, Runciman adds on a different logical level, a consideration of the place of values in research. Recognizing that, by virtue of being human researchers cannot be free of values, he argues for reticence on these matters on the assumption that authentic descriptions of experiences would enable a reader to make his own value judgements. The only value he admits, indeed requires, is goodwill and sympathy *vis-à-vis* the people studied.

Runciman's clarification of the concept of 'understanding' has the advantage of simplicity about the central task of social science; Merton's paradigm has the advantage of greater specificity. The crux of Merton's scheme is the concept of consequences of the phenomenon under study. Following Freud, he distinguishes *manifest consequences* from *latent consequences.* Just as Homer made his audience understand the extraordinary beauty of Helen not by describing it but by the consequences her appearance had on men who swooned away, so Merton suggests that one way of understanding a social phenomenon is to trace its consequences of both kinds, thus adding a fourth sense in which understanding is used in the social sciences.

The other items of the paradigm (here slightly reformulated so as to make their relation to Runciman's scheme terminologically clearer) are: defining and describing the phenomenon under study from the observer's point of view (corresponding to Runciman's reportage); describing subjective dispositions, motives, purposes and experiences from the participant's point of view (corresponding to Runciman's description); manifest and latent consequences for individuals, sub-groups, society, culture (no Runciman equivalent); prerequisites for the phenomenon to occur (vaguely related to Runciman's explanation, a term Merton avoids); social mechanisms such as institutional arrangements, formal and informal norms, etc. (probably subsumed under Runciman's reportage); alternatives, equivalents and substitutes for the phenomenon. Merton here breaks away from an often criticized feature of functionalism, namely the assumption that what is must be so (no Runciman equivalent); the social context of the phenomenon (a specification of Runciman's reportage); dynamics and change (no explicit equivalent in Runciman's scheme); validation of all previous items (Runciman mentions validity only as a criterion for explanation); ideological aspects in research (corresponds to Runciman's evaluation).

Each scheme has some drawbacks. While Runciman's guide to the major formal questions to be asked in research—'what?', 'why?', 'how does it feel?'—seems incomplete,[7] Merton's is less neat; it has proved useful as an ordering device for an otherwise uncoordinated field of research,[8] while revealing at the same time another shortcoming: the scheme is recursive, that is to say it can in its entirety be applied to the data obtained under each separate item. If stigmatization, for example, is one social mechanism operating with regard to the unemployed, it can itself become the phenomenon under investigation to which all items of the paradigm can be applied. This invites, if not infinite regress, then one large enough to limit its ordering capacity.

On the assumption that in combination these two schemes contain the major formal issues that have been tackled by social research, the question arises as to which of them and in what combination have actually been dealt with in social psychological research on unemployment. The total reservoir of strategic questions forms the yardstick with which the collective effort in this field can be assessed.

Strategies in social psychological research on unemployment

From the perspective of macroeconomics, social psychological studies of unemployment deal with its latent, unintended consequences. From the perspective of social psychology, however, the definition of the subject under study is subtly changed. Not unemployment as a phenomenon in the economic order, but the state of being unemployed, is under investigation. The whole set of formal questions can, of course, be applied to this reformulated topic. (Parenthetically it should be noted that the recursiveness of Merton's paradigm which proved to be a difficulty when using it as an ordering device emerges here as an asset.) That one and the same set of observations can be regarded as latent consequences from the viewpoint of one discipline while it presents the definition of the situation from that of another discipline has implications for the logic of interdisciplinary procedure. But more of this later.

A whole host of studies concentrate on reportage and description in various sub-groups defined by their locations, age, sex, previous occupation, social status (minority groups) and length of unemployment. Much has been learned from such research, and there is scope here for many more studies. Two criticisms have, however, been raised about much of this work; one has to do with the adequacy of reportage, the other with the authenticity of the description.

Several investigators[9] have recently pointed out that not all is well with reportage in these studies. The critics maintain that there is often a tacit assumption that unemployment is a psychologically destructive experience, that researchers have therefore concentrated on eliciting negative feelings and that in interpreting results they have ignored individuals for whom unemployment is a welcome relief from unsatisfactory employment and an opportunity to make a new start in their way of living. Such studies can also be criticized in terms of the last item in both formal schemes, the place of value, for having permitted a preconceived position about unemployment to interfere with the adequacy of their reportage.

It is true that most studies have documented negative psychological experiences among the large minority of the unemployed; the massive evidence for this finding seems to me beyond doubt. But it is also true that some of the jobless make constructive use of their enforced free time and feel less deprived, if at all, than others. The temptation to ignore this minority in the interpretation of results is highest in studies involving large numbers where simple statistical averages are used in reporting degrees of well-being. In such studies the averages are consistently lower than for employed comparison groups. Here, as so often with hypothesis-confirming average measures, an indisputable result tells only a major part, not the full truth about a human situation.

Such studies as a rule rely entirely on self-report of feelings and experiences, and this gives rise to the second criticism: are these descriptions authentic? To be interviewed or to complete a questionnaire is a rather unusual event in most people's lives, inviting reflection about the presentation of oneself to the world. The importance of this reflective view is not diminished by pointing out that it may differ from the spontaneous experience of living. If this is granted, the question arises as to whether interview studies can indeed provide a full and authentic account of what it is like to be unemployed. What is required in addition are unobtrusive measures that transcend the special situation with its demand for reflection and the possibility that it may induce the unemployed to manipulate consciously or unconsciously the impression on the interviewer.[10] In the thirties it was possible in at least one instance to provide an account less open to doubt because the observers lived for a period in a community virtually totally unemployed. Records of the change in consumption patterns from the local shop, from the public library, from attendance at voluntary organizations, etc. provided unobtrusive evidence on what it was like to be unemployed; in the particular instance, supporting subjective negative accounts from interviews.[11] Few such efforts are currently being made, even though there are now some communities where unemployment is the rule, no longer the exception. Then it emerged that the unstructured, unlimited time at the disposal of the unemployed was not experienced as leisure but as a burden, and former leisure activities were curtailed. In contrast to that early study, one modern investigation on the reading habits of the unemployed could not confirm a decline in book borrowing.[12] Much more effort and imagination are needed to complement interview and questionnaire studies in this direction.

Reportage and description of subjective states are the basic strategies in social psychological research on unemployment, the strategies in terms of which results are as a rule reported. This is in sharp contrast to experimental psychology which limits itself to reportage and explanation. It is, perhaps, due to the great influence of experimental on social psychology that studies aiming at description are often limited to what people are prepared to say and to inadequate reportage, not only in terms of the first criticism mentioned above but also in ignoring Merton's elaboration of reportage to include social mechanisms and social context.

In this enlarged sense reportage has some bearing on the search for explanations. If the unemployed themselves explain feeling depressed or cut off from wider social contacts by the lack of a job their words must be taken seriously. None the less, such attribution of causality may simplify a more complex causal picture. Studies made during the Great Depression of the thirties document that the unemployed then lived under conditions of abject poverty. Their resignation, apathy and hopelessness may have been due not to the loss of jobs but to their poverty. By and large the unemployed of the eighties suffer from relative, not absolute, economic deprivation. To the extent that they too attribute their psychological state to being without a job they lend some support, though not enough, to the idea that this is a sufficient explanation for their subjective state. To clinch the point, studies of economically privileged unemployed persons whose standard of living is not radically affected by the loss of a job are indicated. Some such studies have been conducted with managers and professional people;[13] once again, a large majority were psychologically disturbed by their unemployment. To be able to say with greater certainty that being without a job rather than relative economic deprivation accounts for subjective negative states may not matter much to the individual unemployed; but it has policy implications and, in the context of this paper, it is relevant for dealing with the 'why?' question.

The search for explanations comes rightly at a late stage in the development of this field of research. Unless other strategic questions have first clarified what it is that has to be explained, the question does not even arise.

In practice, the logically clear distinction between reportage and description on the one hand and explanation on the other is not sharp. Accurate reportage and authentic description already imply some assumptions about possible answers to the 'why?' question. This is so because all research must simplify the full complexity of actual life. The direction of such simplification, that is the decision about what to report and describe, inevitably contains elements of an explanation.

Implicit in several social psychological studies are references to the psychological meaning of employment as an explanation for the experience of being unemployed. So Warr and his research group have demonstrated that those who were much involved with their previous jobs suffered psychologically more when unemployed than those who were not.[14] But even those who left their jobs voluntarily or lost a job they hated show, in their majority, negative responses to unemployment. Degree of involvement is thus an important modifier of experience but not a full explanation. The question then arises as to what employment means for all, even for those who resent the daily drudgery.

For at least two centuries, employment has been a central institution in developed countries. Like every other institution, its rules and organization prescribe for all participants certain unavoidable categories of experience and behaviour. Whether the quality of these experiences is positive or negative, as categories they are inescapable. These categories are: time structure, social contacts beyond family and self-chosen friends, participation in collective efforts, definition of one's social standing and regular activity.[15] Participation in other institutions such as, for example, the church or voluntary organizations also enforces categories of behaviour and experience but these are less entrenched, less regular, less controlled and not linked to the economic necessity of making one's living. The categories enforced by the institution of employment seem to correspond to fairly deep-seated needs of human beings in modern societies: people need some structure to the waking day, they need to enlarge their social horizon beyond their primary groups, they need to be involved in collective efforts, they need to know where they stand in society, and they need to be active. Of course they need other things too, among them the very opposites of the needs met in employment: unstructured time, privacy, individual efforts, self-esteem independent of social norms, and periods of passivity. The overriding need is for complementarity of the two types of need; being employed for sixteen hours a day would create as many psychological disturbances as being unemployed.[16] The notion that the needs met by employment are fairly general is supported by evidence for the negative experiences that many retired people have when they are deprived of the enforced categories and by the sense of emptiness and frustration of many housewives once their children no longer depend on their daily work.

The existence of these needs and the fact that they are traditionally met by employment are thus jointly an explanation for the negative experiences among the majority of the unemployed. By the same token, it also helps to suggest the conditions under which a minority will regard unemployment as a constructive period in their lives: to the extent that out of their own initiative and without institutional enforcement some people meet these needs through other activities, they will suffer psychologically less when unemployed.

This explanation has recently been tested empirically by Ian Miles and found to be valid.[17] He elicited from unemployed men whether or not any of these five needs were met in their current life. Those who had satisfied none were lowest in psychological well-being; those who had met one need came next, and so on in a systematic increase so that those who had met all five needs were highest in well-being, though still below the level of an employed comparison group.

There are then alternatives to employment that can mitigate the predominantly negative experiences of being unemployed. In what manner they can be institutionalized so as to benefit the large majority of young and unskilled people who lack the know-how and initiative to design their own purposeful existence is an important question that has hardly been tackled, though Fryer and Payne have started to think about it.[18]

Another often neglected strategic question is directed to the structural context with its possibilities and constraints in which unemployment occurs. Many social psychologists present their results without going into this matter beyond identifying the locality in which they worked. This is understandable since their competence hardly ever extends to analyzing psychologically relevant features of the social structure, although there is systematic work available from other disciplines that could enrich the interpretation of results. Once again, the issue of interdisciplinarity arises, to be considered below.

Much social research regards explanation as the culminating achievement, in emulation of the natural sciences where this is so often justified. Within the greater complexity of social phenomena, understanding remains incomplete, however, without tracing their consequences. The strategy *par excellence* in social research is the search for latent consequences which, by definition, transcend the description of subjective states, reportage and explanation.

To make visible consequences that remain invisible to the naked eye is a complex task. Several investigators have claimed to discover latent consequences of being unemployed when what they have actually achieved is the discovery of latent facts and events.[19] This is no mean achievement in its own right as long as it is recognized for what it is. Harvey Brenner's macro-statistical studies, for example, about the correlation of rates of unemployment with various indicators of social pathology (morbidity, mortality, suicide, crime, etc.) have led to much controversy simply because of a confusion between latent facts and latent consequences.[20] All macro-statistical work is bound to reveal latent facts for the simple reason that not even the most acute observer can encompass the complexities of large social aggregates. Brenner may well be right about his claim to have discovered consequences of unemployment, but he has not proved it.

To go beyond such plausible suggestions with regard to latent consequences requires coming to grips with the indubitable fact that social events invariably have a multitude of consequences, just as they have a multitude of causes. To make the consequential nature of social pathology more convincing, quasi-experimental methods are in order: if people's state of mind were assessed before they became unemployed and then compared with that after a period of joblessness, the differences, if any, could with greater confidence be identified as consequences. The practical difficulties confronting such longitudinal research on the nationwide scale on which Brenner operated are obvious. On the more modest scale of empirical social psychological research, a few such studies exist. Guerney, for example, had the hunch that unemployed school leavers would suffer in the development of social maturity, as compared to their employed peers.[21] She assessed children in their last school year and again months after the end of school and found that those who then had no job did not show further constructive development while those with jobs did; the two groups had been fairly well matched on the relevant measures at school.

Occasionally, such longitudinal research is facilitated when appropriate records are available for an earlier period. In this way one study discovered as a latent consequence of parental unemployment a deterioration in children's school performance by comparing their marks before and after a father lost his job.[22] A replication of this study from the thirties would not be difficult and is highly indicated.[23]

One final strategic question from Merton's paradigm deserves attention: change and dynamics. There are two time perspectives which need to be considered in investigating change, if any, in the experience of being unemployed: has it changed in historical time spans and does it change in the present period with the duration of individual unemployment?

Both questions have been tackled by social psychologists. With regard to the first, detailed empirical comparisons are, however, handicapped because of the considerable superiority of modern research methods. To the extent that broad comparisons have been made, the contemporary experience of being unemployed shows many similarities with that described during the Great Depression, notwithstanding the enormous social changes since then. Had the early methods been quantitatively more sophisticated so that exact replications could be made now, some differences might perhaps have appeared. As it is, the apparent similarities can be speculatively understood from the fact that the structure of employment as an institution enforcing certain categories of experience has not changed during the last half-century.

One difference has, however, emerged from such comparisons with regard to the second time dimension: changes in experience with the length of individual unemployment. In the thirties several phases could be distinguished: an initial period of shock reaction was followed by a slight recovery which then gradually subsided and led for the majority to resignation or apathy after a period of two or more years.

Current studies, although dealing with somewhat shorter time periods, do not discover progressive deterioration.[24] In the light of available research it is not possible to decide whether or not the early phases are replicated in the present situation. Since the unemployed in the thirties were gradually reduced to subsistence levels, their hopelessness in the last observed phase may well have been the inevitable consequence. For the present, more research on the impact of duration is indicated. It has been demonstrated that the longer a person is without a job, the less is the probability of their finding one. Whether this is due more to employers' assumptions about reasons for long-term unemployment or more to hopelessness or to becoming unemployable through prolonged idleness or to other factors remains to be discovered.

Before leaving the guidance provided by the formal schemes in reviewing strategic questions, the place of values and ideologies in research on unemployment should be briefly explored, even though the questions to be considered here are not strategic in the sense of directing investigators to what should be asked about the phenomenon under study. Rather, they are questions about the social scientists, to be asked by themselves or by others. Runciman's position in this respect has already been presented; it is predicated on the assumption that investigators have clearly articulated values so that conscious decisions to keep them out of their work have a chance of being implemented. Merton is more aware of the subtle ways in which a researcher's social position, source of finance or personal political values can unwittingly introduce bias into problem formulation and other procedures. An example of such possible bias was mentioned earlier when the tacit conviction that unemployment was psychologically 'bad' led to ignoring counter-examples. The issues that Merton raises in this context are beyond simple remedies. For two reasons it is important to keep them in mind: first, as a warning to the consumer of research that value bias may influence results, and second because this recognition supports the demand for similar strategies to be used by different observers in tackling identical research topics. Differences, if any, should make it easier to discover implicit value premises.

The preceding application of the formal strategic questions to the social psychological study of unemployment is not complete, neither with regard to possible combinations of questions nor, of course, with regard to the literature. It should suffice, however, for the identification of dominant trends and gaps.

The dominant strategies in the field appear to be reportage and description. As has been pointed out, this is essential but not enough, not even when descriptions are refined with the help of unobtrusive measures or through the introduction of comparison groups or modifiers such as work involvement. There exists a remarkable German study of unemployed women which demonstrates the dangers as well as the assets of driving a combination of description and reportage to its limits.[25] The authors managed exceptionally well to obtain a large and adequate sample of women. In their analysis of unemployed women they qualify their description by age, length of unemployment, financial situation, nature of previous job, family situation, status of partner, education, skill level, work motivation, political interests, perception of causes of unemployment, and more. Unquestionably, these factors do modify the experience; by the same token, however, such abundance of qualifications makes it well-nigh impossible to see the wood for the trees. Of course, the experience of every unemployed person is unique; but to document this is hardly the task of research. What one does learn from this and other studies is that the unemployed are not a 'group' in any psychologically legitimate sense of that term; only in having lost a job and thus being different from the accepted norm of society are they all alike.

Many studies are therefore restricted to samples of the unemployed who have more than the absence of a job in common—school leavers, occupational groups, etc. While this limits the number of possible independent variables it raises the issue of generalizability of results. Kelvin and Jarrett note the tendency of social scientists to generalize from such studies to all unemployed, and to warn against it.[26] Generalizations, they maintain, are adopted as popular stereotypes to which the unemployed themselves are exposed and may thereby acquire the status of self-fulfilling prophecies so that the unemployed begin to feel stigmatized, for example, because everybody expects them to feel so. None the less, many investigators rightly aim for generalizations and find a legitimate way of doing so by combining reportage and description with an explanation that transcends the concrete circumstances of an isolated study and that can therefore be applied to another concrete situation.

As has been pointed out before, there is in the social psychological research literature little evidence of elaborated reportage, though the inclusion of concern with the social context may be an appropriate way of discovering latent consequences for the wider community of how millions of unemployed people feel and think. Brenner's attempts in that direction remain controversial because he established correlated facts, not consequences. Even though this criticism is somewhat mitigated by his introduction of time lags between unemployment rates and indicators of social pathology, there remains another objection: he used nation-wide indicators which of course include the large majority of the employed population whose share in social pathology may also have changed in either direction for whatever reason. It is indeed immensely difficult to conclusively trace the latent consequences of mass unemployment for society at large. And yet there can be little doubt that such consequences exist.

Speculatively, several such consequences have been suggested. The growth of the informal economy, particularly the black economy, has been said to be the consequence of large-scale unemployment. There is, however, no empirical evidence to support this while the documented lack of skills, lack of sophisticated tools and the often depressed and passive mood among the unemployed are important counter-arguments.

Another suggestion blames mass unemployment for the increase in crime, but the evidence is inconclusive and beset by all the well-known vicissitudes of crime statistics. Others attribute to mass unemployment the lack of consensus in society; the nation is now split in two, so it is maintained, not according to social class but to employment status. Once again there is no empirical evidence to support this. Perhaps it is beyond the competence of social psychology to provide evidence for such macro latent consequences. It should be stressed, however, that all these speculations are predicated on social psychological common sense assumptions. To identify them and discuss their probable validity in the light of established social psychology knowledge is an urgent and more manageable task.

Some social psychological research takes it for granted that being without a job is a sufficient explanation for the feelings and moods among the unemployed. Two qualifications of this simple causality have been proposed. First, since unemployment is as a rule coupled with a significant drop in the standard of living, relative poverty, not the lack of a job, may fully account for the state of mind. Second, the experience of being unemployed is not uniform but mediated by individual predispositions, life histories and circumstances. To take either of these alternatives in isolation implies shifting from a social psychological perspective to a sociological one in the case of relative deprivation, to a psychological one in the case of individual differences, with their implied models of man as either determined by social conditions or as the maker of his own fate. But the entire *raison d'être* of social psychology lies in its effort to combine both models. This is not to deny, of course, that for many purposes the search for causality has made important contributions to knowledge. Social psychology—so far perhaps with less success than other social sciences—however, seeks explanations that combine social and individual determinants in continuous and simultaneous interaction, neither one preceding the other, as if they were independent and dependent variables. One such explanation has been suggested, but it needs further testing, clarification, modification and perhaps radical change. It is at least social psychological in character, however, linking deep-rooted individual needs to the tradition of having them met by a social institution—employment.

It would be tedious to elaborate further the implementation or the failure to implement all the strategies suggested by the formal schemes. The point of the preceding discussion is to show the utility of using these formal schemes to raise questions about the collective effort in this research field in the hope of perhaps stimulating some to extend the dominant trend in question-asking.

Interdisciplinary research on unemployment

Interdisciplinarity has for the last few decades been a much favoured slogan among social scientists (though not necessarily in universities or funding organizations). There are good and obvious reasons for the implicit recognition that no single discipline can adequately deal with a social phenomenon such as unemployment. Economists, political scientists, sociologists, statisticians, historians, clinical psychologists, students of social administration and of public health, in addition to social psychologists, have relevant contributions to make and have indeed made them. And yet, interdisciplinary research on unemployment has remained a slogan that seems to defy implementation. I do not know of a single study that has advanced knowledge in more than one of these disciplines. Why?

Social psychology, itself a hybrid, demonstrates the difficulty. Most social psychological studies and theories are concerned either with social regularities based on common sense psychological assumptions or with psychological regularities based on common-sense sociological assumptions, not with both. This is so, not just because individual investigators have their primary commitment to one or the other discipline, although this plays a role. The major reason is less easily rectified. The existing disciplines in the social sciences have gradually emerged from an all-embracing philosophy of previous periods by defining specific points of view from which to study the bewildering complexity of the social world. In doing so they have lost in scope and gained in depth and precision of thought. They have chosen different units of observation for which different concepts and methods are appropriate and which give rise to theories that can be tested, changed or abandoned, only by keeping within the same universe of discourse. The degree of specialization achieved by every one of these legitimate points of view makes their combination in research immensely difficult, if not impossible. If neighbouring fields, such as sociology and psychology, have not yet achieved a common procedure, the chances for an integration of less related fields are slim.

This unsatisfactory state of affairs has given rise to a reformulation of the demand for interdisciplinarity. Recognizing that problems in the social world can be tackled by various disciplines in various ways but not simultaneously by various disciplines in the same research process because they ask different questions and use different units, concepts and methods, the demand for interdisciplinarity is gradually being replaced by a demand for multidisciplinary approaches by multidisciplinary teams, every one of whose members is operating within a single discipline. Integration of the resulting diverse findings remains an art that has so far defied codification, although it has a number of outstanding practitioners, among them Christopher Freeman who is fully aware of the 'blinkered view of a single discipline'.[27]

As has been pointed out before, in the case of unemployment the definition of the topic for social psychology can be construed as a latent consequence of unemployment when viewed by economists, even though there are also latent consequences within an economic definition, for example the skill composition of the labour-force. Economists can deal with the latter; social psychologists are required for the former. This is why there are no global theories of unemployment, only partial ones, a fact that is not easily accommodated by those for whom theory is the ultimate goal of science, rather than a tool for the advancement of knowledge.

If I am right in the foregoing remarks, they throw some light on the well-known dilemma in economics about how to combine the differences between macro and microeconomic results. By any reasonable definition of disciplines, microeconomics is *de facto* social psychology, not economics; it deals with behaviour of individuals in their economic context, not with economic regularities. Traditional definitions of disciplines have emerged by historical accidents and result in the fact that some so-called 'microeconomic' studies that could be proudly claimed by social psychology present an unintegrated dilemma for economics.

Given the need for a multidisciplinary approach to the study of unemployment, it goes without saying that the foregoing discussion from the point of view of social psychology is inevitably only part of the story. Subsequent chapters should enable the reader to enlarge the picture, while also demonstrating both the need for and the difficulty of achieving a multidisciplinary interpretation.

Notes

1. See, e.g. K. Macky and H. Haines, 'The Psychological Effects of Unemployment: A Review of the Literature', *New Zealand Journal of Industrial Relations*, **7** (1982), pp. 123–5; or R. M. Guerney and K. Taylor,'Research on Unemployment: Defects, Neglect and Prospects', *Bulletin of the British Psychological Society*, **34** (1981), pp. 205–13.
2. Peter Kelvin and Joanna Jarrett, *Unemployment, its Social Psychological Effects*: European Monographs in Social Psychology, Cambridge, Cambridge University Press, 1983, pp. 2–3.
3. Robert K. Merton, *Social Theory and Social Structure*, Glencoe, Ill., The Free Press, 1949.
4. W. G. Runciman, *A Treatise on Social Theory*, Vol. 1: *The Methodology of Social Theory*, Cambridge, Cambridge University Press, 1983.
5. Merton, op. cit., note 3, p. 21.
6. The brief exposition of Runciman's and Merton's schemes is tailored to the purposes of this article; it cannot convey their rich conceptual content.
7. Runciman has promised to demonstrate in two further volumes the theoretical and practical usefulness of his guide. It remains to be seen whether and how he will deal with matters apparently missing from his guide in Volume 1.
8. Marie Jahoda and Howard Rush, *Work, Employment and Unemployment: An Overview of Ideas and Research Results in the Social Science Literature*, SPRU Occasional Paper Series, No. 12, Brighton, Science Policy Research Unit, 1980.
9. See, e.g., Stephen Fineman, *White Collar Unemployment: Impact and Stress*, Chichester, John Wiley, 1982; or Jean Hartley and David Fryer,'The Psychology of Unemployment: A Critical Appraisal' in G. Stephenson and J. Davids (eds), *Progress in Applied Social Psychology*, Vol. 2, Chichester, John Wiley, 1984.
10. E. J. Webb, D. T. Campbell, R. D. Schwartz and L. Sechrest, *Unobtrusive Measures: Nonreactive Measures in the Social Sciences*, Chicago, Rand McNally, 1966.
11. Marie Jahoda, Paul F. Lazarsfeld and Hans Zeisel, *Die Arbeitslosen von Marienthal*, Leipzig, Hirzel, 1933. English translation, London, Tavistock, 1972.
12. David Fryer and R. L. Payne, 'Book borrowing and unemployment', *Library Review*, **32** (1984), pp. 196–206.
13. e.g. Fineman, op. cit., note 9.
14. Peter B. Warr is director of the MRC/SSRC Social and Applied Psychology Unit at the University of Sheffield. This unit has conducted a large number of studies on the impact of unemployment. Peter Warr has summarized their findings in a paper entitled 'Work, Jobs and Unemployment', available from the unit.
15. I have developed these ideas more fully in *Employment and Unemployment: A Social Psychological Analysis*, Cambridge, Cambridge University Press, 1982.
16. The current balance between employment and leisure based on an eight-hour working day is, of course, not sacrosanct. The needs met by employment would also be met by a working day of, say, six hours. This has obvious implications for dealing with mass unemployment; obvious, but very difficult to implement.

17. Ian Miles, *Adaptation to Unemployment?*, SPRU Occasional Paper Series, No. 20, Brighton, Science Policy Research Unit, 1983.
18. D. M. Fryer and R. L. Payne, 'Proactivity in unemployment: findings and implications', *Leisure Studies*, **3** (1984), pp. 273–95.
19. Bernard Bailyn, 'The Challenge of Modern Historiography', *The American Historical Review*, **87** (1982), pp. 1–24. In this presidential address Bailyn has made the discovery of latent events a key concept in his discussion of historiography.
20. M. Harvey Brenner, *Estimating the Social Costs of National Economic Policy: Implications for Mental and Physical Health, and Criminal Aggression*, Joint Economic Committee of Congress, Paper No. 5, Washington DC, US Government Printing Office, 1976. This is Brenner's most substantial and influential publication. He has written many papers since then, supporting his original conclusions and answering several attacks on his methodology that appear to have an ideological undertone.
21. R. M. Guerney, 'The Effects of Unemployment on the Psycho-Social Development of School-Leavers', *Journal of Occupational Psychology*, **53** (1980), pp. 205–13.
22. Quoted in P. Eisenberg and P. F. Lazarsfeld, 'The Psychological Effect of Unemployment', *Psychological Bulletin*, **35** (1938), pp. 358–89.
23. The relative neglect of this research area emerges from a recent review of relevant studies: Nicola Madge, 'Unemployment and its Effects on Children', *J. Child Psychol. Psychiat*, **24**, No. 2 (1983), pp. 311–19. The author begins her conclusion thus: 'The evidence is limited and inconclusive . . .'.
24. See Fineman, op. cit., note 9; Warr, op. cit., note 14; Miles, op. cit., note 17.
25. K. Heinemann, P. Röhrig and R. Stadie, *Arbeitslose Frauen im Spannungsfeld von Erwerbstätigkeit und Hausfrauenrolle*, 2 vols., Melle, Ernst Knoll, 1980.
26. Kelvin and Jarrett, op. cit., note 2.
27. Christopher Freeman, 'Technological Change and the New Economic Context' in S. Hill and R. Johnston (eds), *In Future Tense? Technology in Australia*, St. Lucia, University of Queensland Press, 1983, p. 55.

8 Technical change and fluctuations in employment

Bruce Williams

In the second decade of the nineteenth century, groups of English artisans destroyed new textile machinery in the Midlands and North of England on the grounds that it took away their livelihood. The effects of the labour-saving machinery on employment were debated by Malthus and Ricardo. At first Ricardo argued that the introduction of machinery could not lead to a permanent displacement of labour, but later—in the third edition of his *Principles*, 1821—he decided that in certain circumstances it would. However, his arithmetical example of permanent displacement was not a simplification but a distortion of reality and had very little influence on economists. Marx was the first economist to treat technical change as a powerful endogenous force which generated fluctuations in employment as well as growth in wealth. In 1930 Keynes wrote an important essay on technological unemployment arising from a temporary imbalance between discoveries of the means of economizing the use of labour and discoveries of new uses for labour. But during the long period of high rates of growth and unusually small fluctuations in employment after the Second World War, there was a disposition to believe that high levels of expenditure on applied research and development and government measures to sustain aggregate demand would prevent significant fluctuations in employment. However, the decline in growth rates and employment from the mid-seventies revived interest in the issues debated by Malthus, Ricardo, Marx, Kondratieff and Keynes in earlier depressions.

A brief account of the way economists have responded to (or in some measure ignored) contemporary events may contribute to a better understanding of the present depression and of the issues which call for further investigation.

Expectations of unemployment

That there is a persistent fear in all industrial societies that the introduction of new technologies will create unemployment is not surprising. Some types of employment have been eliminated and others greatly reduced by changes in technology. In agriculture, for example, where demand for the product is still increasing, employment has been greatly reduced. Employment in agriculture in Britain is now less than 3 per cent of the labour-force, whereas 150 years ago it was 40 per cent. In the United States, which is a net exporter of foodstuffs, employment has fallen from 40 per cent to less than 4 per cent in the past hundred years, while in Japan the reduction has been from about 70 per cent to 10 per cent. Such drastic reductions in the proportion of the labour-force in agriculture are the consequence of the invention and adoption of better seeds, breeds, farming practices and pesticides which increase yields per acre, and the invention and adoption of labour-saving devices such as tractors, combine harvesters, milking machines and herbicides.

The invention of possible new processes of production such as continuous casting or electronic capital equipment, and the use of these inventions in the production system, frequently reduce the labour needed for each unit of output, though the consequences for employment depend on the extent of the increase in output. Inventions also make possible the introduction of new products which extend the demand for labour. The prelude to the industrial revolution in Britain in the second half of the eighteenth century was the agricultural revolution of the first half. That revolution released labour which was used to increase the production of traditional manufactures, and then to increase the range of goods and services. Later, despite that great increase in the range of products, the proportion of the workers employed in industry also fell. In Britain the proportion of workers engaged in industry rose from 30 per cent in 1820 to about 45 per cent in 1890 and then came down to 38 per cent by 1980. The proportion in services rose from 30 per cent in 1820 to 40 per cent in 1890 and then to almost 60 per cent in 1980. There have been similar changes in the structure of employment in other industrial countries.[1]

The growth of employment in the service sector has been one of the most striking features of industrial evolution. However, there are now fears that the growth potential of the service sector will not be sufficient to absorb the labour released from industry as a consequence of advances in robotics and electronic control systems. Indeed, there are fears that employment in services will cease to increase, and even that the further improvement and adoption of information technology will lead to a reduction in employment in the service sector.

If the proportion of the labour-force employed in each sector falls, the proportion of the labour-force in work must fall. The fears that the opportunities for employment in industry will continue to contract and that the spread of information technology will also reduce employment in the service sector are pictured in Figure 8.1. Employment in the sectors is expressed as a percentage of the labour-force and not, as is customary, as a percentage of employment. Unemployment between 1700 and 1980 is given as 5 per cent, which was the average level. The possible movement

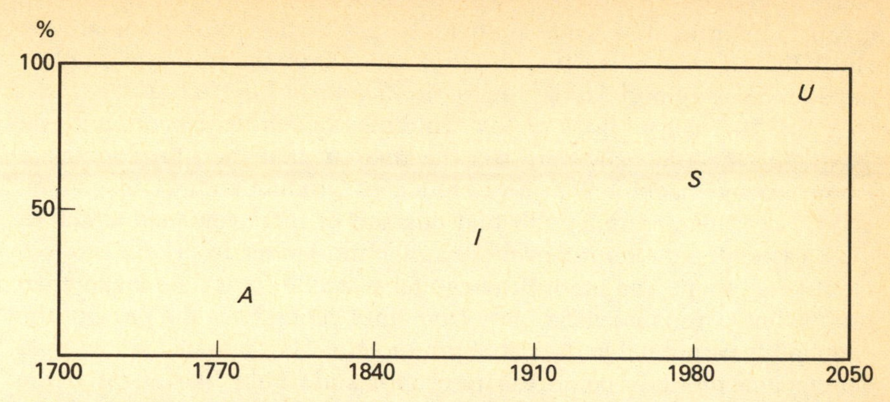

Note: *A* is the percentage of the labour-force employed in agriculture, *I* and *S* the percentages in industry and services, and *U* is unemployment.

Figure 8.1 Changes in the structure of employment to 1980 and possible changes in the future (at current life hours of work)

in unemployment shown in the figure implies something like the current level of hours of work.

The classical economists

In *An Inquiry into the Nature and Causes of the Wealth of Nations* (1776), Adam Smith emphasized the role of the division of labour in improving the productive powers of labour: 'the invention of all those machines by which labour is so much facilitated and abridged, seems to have been originally owing to the division of labour'. As men specialize in particular operations, they invent easier and readier methods of performing them; as the making of machines becomes a specialized occupation, the makers improve machinery; as science becomes the principal or sole occupation of a particular class of citizens, the latter too improve or invent machines. 'It is the great multiplication of the production of all the different arts, in consequence of the division of labour, which occasions in a well governed society, that universal opulence which extends itself to the lowest ranks of the people.'[2] In a free enterprise system the market mechanisms would bring supplies and demands into equilibrium in all markets.

In *Principles of Political Economy and Taxation* (1817) Ricardo further developed the doctrine that in the absence of restraints on trade the price system would equate supplies and demands and ensure full employment. But Malthus, the author of the *Essay on Population* (1798), was not convinced that it was impossible for the demand for labour to be deficient, and wrote that 'an attempt to accumulate very rapidly . . . by greatly impairing the usual motives to production must prematurely check the progress of wealth . . . and consequently the power of employing an increasing population.' (Letter to Ricardo, 7 July 1821.)[3]

In the first two editions of his *Principles of Political Economy and Taxation*, Ricardo had expressed the opinion that the introduction of labour-saving machinery to any branch of production was in general good, 'accompanied only with that position of inconvenience which in most cases attends the removal of capital and labour from one employment to another'. The innovating capitalist would for a time make great profits, but as the new machinery came into general use the price of the commodity would fall to its cost of production. The capitalist would then get the same profits as before the innovation and would, like the rest of the community, get the benefits of the innovation in the form of 'a greater quantity of comforts and enjoyment'. But then in the third edition of his *Principles*, published in 1821, Ricardo wrote that he had become 'convinced that the substitution of machinery for human labour is often very injurious to the interests of the class of labourers'.[4]

Ricardo wrote that he had been mistaken to suppose that whenever the net income of society increased its gross income would also increase. To explain his error, he assumed an initial capital of £20,000, of which £7,000 was fixed and £13,000 circulating. With a profit of 10 per cent on the £20,000 capital employed, the net produce would be £2,000 and gross produce £15,000 (the £2,000 plus the £13,000). In the second situation he assumed the capitalist to employ his £20,000 very differently—that he would use half his men in constructing a machine worth £7,500, so raising fixed capital to £14,500 and reducing circulating capital to £5,500. Net produce (the 10 per cent on capital employed) would stay at £2,000 but gross produce would come down to £7,500. With the circulating capital available for the employment of labour down from £13,000 to £5,500 distress and poverty would follow.

But Ricardo did not end up with the conclusion that this could justify opposition to the introduction of machinery. Having proved to his satisfaction that 'the discovery and use of machinery might be followed by a diminution of gross produce' (p. 266), he proceeded to conclusions which were at odds with his starting position that the substitution of machinery for human labour is often very injurious to the interests of labourers. He did so on two grounds. The first ground was that the reduction in commodity prices that followed the introduction of machinery would add to 'the efficiency of the net revenue'—that is, add to the real value of profits—and so increase savings. This would add to circulating capital and the employment of labour. The second ground was that to elucidate the principle he had supposed a sudden discovery of new machinery, 'but the truth is that these discoveries are gradual and rather operate in determining the employment of capital which is saved and accumulated than in diverting capital from its actual employment' (p. 270).

But even if in market economies departures from full employment were soon corrected, so that all classes shared in the increase in produce made possible by more and more advanced machinery, it does not follow that all classes would share equally. Suppose, for example, that the capital stock is 300, the output 100, and the equilibrium gross profit 10 per cent of capital. The share of capital in output would be 30 and of labour 70. Suppose next that inventions raised the average rate of profit and led to an increase in the capital stock by 100 per cent to 600, at which level the gross profit became 10 per cent again. If the output also increased by 100 per cent, to 200, labour would receive 140, and its share would remain at 70 per cent. But if the inventions raised the capital/output ratio from 3 : 1 to 4 : 1, implying a rise of output from 100 to only 150, owners of capital would receive 60 (10 percent of the new capital stock of 600), and labour the remaining 90. The income of labourers would rise from 70 to 90 but their share would fall from 70 to 60 per cent. If, however, the inventions reduced the capital/output ratio to 2.5 : 1, implying a rise in output from 100 to 240, owners of capital would still receive 60 (10 per cent of the new capital stock), but labour would receive the remaining 180. Labour's share would rise from the original 70 to 75 per cent.

Capital/output ratios have varied over time. In the United States, between 1919 and 1957, the ratio fell substantially. It fell so much that, despite a rise in the rate of profit by 10 per cent, the share of labour rose from 72 per cent to 80 per cent. Capital/output ratios also change during periods of boom and slump. In the first example above, the 'equilibrium' capital/output ratio of 3 : 1 was followed by another 'equilibrium position' where both capital and output had been increased by a factor of two, leaving the capital/output ratio unchanged. But during the move from one equilibrium position to another, the capital/output ratio would rise. Thus, when capital had reached 450, the time taken to commission new plant would have prevented output from rising above, say, 130. The capital/output ratio would be 3 : 5 at that stage of movement to the new equilibrium position. And if, as has frequently happened in the past, the increase in capital overshot the equilibrium position by, say, 5 per cent, the capital/output ratio as normally measured would rise. It would rise still more if, as a result of the excess capacity, output did not rise to one-third of 600 but to 175 only. The capital/output ratio would then be 3.6 : 1, until employment and output revived and any surplus capital was written off. But such an increase in the capital/output ratio due to 'overshoot' would not lead to an increase in capital's share but to a reduction. Labour's share would rise because of a sharp fall in the rate of profit, but it would be a larger share of an output which was less than the full-employment level.

Problems of distributive shares will seem less important when output is increasing substantially than when it is increasing slowly. To the followers of Malthus and Ricardo, the law of diminishing returns and the tendency to a stationary state were central preoccupations. Had such economists foreseen the extent of change that invention and capital accumulation would make possible, they would probably have given the theory of distribution a less prominent place in their thoughts. Even J. S. Mill in his *Principles of Political Economy*, written in 1840, fifty years after the publication of the first *Essay on Population* by Malthus and thirty years after the publication of Ricardo's *Principles*, included a chapter on 'The Tendency to a Stationary State'.[5] Yet, in an earlier chapter, he had written that physical knowledge was advancing more rapidly and in a greater number of directions than in any previous age or generation, and was 'converted by practical ingenuity into physical power' more rapidly than at any previous period.[6] He added that 'the manual part of these great scientific operations is now never wanting to the intellectual', and came near to implying that the combination of advances in knowledge and practical ingenuity could overcome the limits to growth. But he did not in that context consider the possibility of technological unemployment.

Marx and technological unemployment

Marx did not accept the Malthusian and Ricardian doctrine of the limits to growth. His view, more in line with that of Adam Smith, was that productive division of labour goes on increasing with the extent of the market. In the *Communist Manifesto*, published in the same year as Mill's *Principles*, Marx and Engels wrote that 'meantime the market kept ever growing, the demand ever rising. Even manufacturing no longer sufficed. Thereupon steam and machinery revolutionized industrial production. . . . The bourgeoisie cannot live without revolutionizing production.'[7] In 1857/8 Marx wrote that when available machinery already provided great capabilities and large industry had already reached a higher stage and all the sciences had been pressed into the service of capital, then 'the analysis and application of mechanical and chemical laws enabled machines to perform the same labour as that previously performed by the worker. Invention then becomes a business, and the application of science to direct production becomes a prospect which determines and solicits it.'[8]

In the *Communist Manifesto* Marx and Engels wrote that 'the essential condition for the existence and for the sway of the bourgeois class is the formation and augmentation of capital' (p. 93), or as Marx put it in *Capital*, 'accumulate, accumulate! That is the Moses and the Prophets'. In *Capital* Marx developed a theory of fluctuations determined by capital accumulation and technical change: normally there would be a 'reserve army' of unemployed which would serve to keep wages down, but periodically wages would rise as the stock of capital which determined the employment of labour caught up with the supply of labour. This rise in wages would reduce profits and thereby check the accumulation of capital. To restore profits there would be labour-saving innovations which would create technological unemployment. Wages would then fall, profits would rise, the process of capital accumulation would be resumed, the stock of capital would grow relative to the supply of labour, and the size of the industrial reserve army would contract.

To his analysis of past fluctuations in the industrial reserve army, Marx added a prediction that these fluctuations would grow in magnitude as the evolution of technology both raised fixed capital relative to working capital and so generated an underlying tendency for the rate of profit and therefore the rate of accumulation to fall, and provided cumulative economies of scale in production leading to stronger pressures to adopt more labour-displacing equipment, increase the industrial reserve army and reduce wage costs.

In Marx's description of the process of production, the use of the initial fixed capital (c) and variable capital (v) creates surplus value (s). The rate of surplus value is s/v, and if that is constant the predicted rise in s/v must result in a fall in the rate of profit—$s/c+v$). But if s/v is constant, real wages must increase with increases in productivity. It is therefore not surprising that Marx's longer-term theory of capital evolution—or dissolution—has not fitted the course of events. Real wages have increased, and although economies of scale in some fields of production have resulted in monopolies or oligopolies the predicted tendency to progressive monopolization and labour displacement has been checked by developments in science and technology which have created opportunities for new firms, new products and new types of employment. In his path-breaking analysis of the interaction between changes in technology and production, Marx concentrated too much on process innovations in established products, and underestimated the significance of product innovations. His analysis of the role of technical change in generating short-term fluctuations in employment was sounder than his theory of long-term trends in wages, profits and technological unemployment.

Long-cycle theories

Industrialization increased the regularity and probably the extent of fluctuations in economic activity. In the nineteenth century there was an increase in the proportion of the population affected by trade cycles; but there is no clear evidence of a trend increase in the amplitude of cyclical fluctuations in employment. Statistics of the trade-union unemployed in Britain from the eighteen-fifties indicate that average unemployment was 5 per cent in the fifties and sixties, a little under 4 per cent in the seventies, just over 5.5 per cent in the eighties and just under 4.5 per cent in the nineties. From 1870 to 1913 peak unemployment was 11.4 per cent in 1879, 10.2 per cent in 1886, 7.5 per cent in 1893 and 7.8 per cent in 1908.[9] Nor is there a clear sign of a trend in Maddison's estimates of the growth rate of output between 1820 and 1870.

After 1873 there was a fall in consumer prices in almost all industrial countries. Between 1873 and 1895 consumer prices in Britain and the United States fell by 30 per cent, in Germany by 12 per cent and in France by 10 per cent. Prices then rose again up to the outbreak of the First World War. In Britain—though not in the United States, France and Germany—there was also a fall in growth rates between 1870 and 1913.[10]

In 1913 van Gelderen and Pareto drew attention to the signs of something like fifty-year cycles in the movements of prices and interest rates. Then, in 1926, Kondratieff, Director of the Business Cycle Research Institute in Moscow, argued from the movements since 1779 in wholesale prices, wages, interest rates, bank deposits, the values of foreign trade, coal production in England, coal consumption in France, and pig iron and lead production in England, that there were signs of long cycles of forty to sixty years, medium cycles of seven to ten years, and short cycles of three to four years.[11] It is not possible to draw any significant conclusions about fluctuations in output and employment from the physical indicators used by Kondratieff. His main evidence for the existence of long waves comes from the interrelated monetary indicators as seen earlier. It was not until the twenties in Britain and the thirties more generally that there was a substantial rise in unemployment in the down-swing of a period which Kondratieff and followers identified as the down-swings of long waves.

In 1930, Kuznets published the results of his more systematic analysis of a larger number of price and output series and concluded that there had been secular variations in production similar in most cases to those in prices, that the complete swing was only twenty-two years for production and twenty-three for prices, but that the evidence would not support a conclusion that these variations were cycles.[12] Despite that, Schumpeter later wrote of fifty-year long cycles, with four stages of prosperity, recession, depression and revival, as if the existence of regular long cycles had been proven.[13]

The main significance of Schumpeter's work on cycles was the extension of his earlier work on the role of entrepreneurs in economic development, and the identification of the role of particular new technologies.[14] His starting-point was the tendency of competition to eliminate profits and the possibility of re-creating profits by innovations. The profits of successful investments in innovation would encourage imitations and improvers and substantial further investment activity. The profits of innovation would then be competed away and the investment boom would tail off. During the subsequent depression the least efficient firms and equipment would be eliminated, the new processes and products would be improved and consolidated, and the economic system would begin to settle down at a higher level of output per head—until disturbed by new entrepreneurial activity.

In Schumpeter's account of the introduction and diffusion of new technologies, there is much less emphasis on fluctuations in employment than in Marx, which is rather surprising for an account completed after the severe depression of the nineteen-thirties. There was a substantial increase in unemployment between the two world wars. In Britain the average level of unemployment rose from less than 5 per cent between 1900 and 1914 to 9.5 per cent in the twenties and 11.5 per cent in the thirties. In the United States unemployment rose from less than 5 per cent in the twenties to over 14 per cent in the thirties.[15] Schumpeter treated the Great Depression of the nineteen-thirties in a very cavalier and unhistorical fashion: 'the depression that ran its course from the last quarter of 1929 to the third quarter of 1932 does not prove that a secular break had occurred in the propelling mechanism of capitalist production because depressions of such severity have repeatedly occurred—roughly once every fifty-five years'.[16] But they had not. Unemployment was much more severe, and the extent of the fluctuations in output much greater, than in earlier depressions.

Technology and fluctuations since 1913

The differences between movements in prices, production and productivity in the hundred years that followed the end of the Napoleonic wars and the seventy years since the beginning of the First World War are much greater than the similarities. Since 1913 there have been two world wars, a marked acceleration in the growth of population, and substantial increases in the proportion of resources devoted to invention and innovation, in the average rate of growth in output per head and in the range of fluctuations in employment and prices.

In the hundred years after 1750, world population increased by 60 per cent and in the hundred years after 1850 by 130 per cent. It then increased by 80 per cent in the next thirty years. That increase was made possible by successful experiments in plant and animal breeding, by improvements in the knowledge of soils and the conditions of plant growth, and by innovations in the production of agricultural machinery, fertilizers and pesticides. Organized research and development activities in the agricultural, mechanical and chemical industries have played an increasingly important part since the eighteen-seventies in transforming agricultural productivity.

In expounding the role of division of labour in productivity, Adam Smith had referred to the emergence of science and invention as specialized activities, and Marx wrote of invention as a branch of business once heavy industry had reached an advanced stage and the various sciences had been pressed into the service of capital. But although the major growth in organized industrial research and development did not take place until after the Second World War,[17] it had gone far enough by the late nineteen-twenties for Schumpeter to write that planned inventions were making large corporations the main vehicles for technological innovation.[18]

Between 1870 and 1913, the annual growth in output per head of population in the industrialized countries averaged 1.4 per cent. Between 1913 and 1950—a period which included two world wars and a deeper depression than any in the previous century—the annual growth averaged only 1.2 per cent (though the growth in output per man-hour rose from 1.6 per cent to 1.8 per cent). In that period average unemployment was considerably greater than previously. In Britain unemployment averaged 9.5 per cent in the twenties, 11.5 per cent in the thirties, and almost 15 per cent in 1930–32. In the United States, where the growth in productivity remained high until the thirties, unemployment averaged less than 5 per cent in the twenties, but rose to over 14 per cent in the thirties and to almost 20 per cent in 1930–32.

In Britain and the United States between 1880 and 1895, consumer prices had fallen by one-third. There was a rather greater fall between 1920 and 1933 in Britain, but not in the United States, though apart from the sharp inflation at the end of the war the fall in prices was more restrained. Between 1922 and the depth of the Depression in 1933 the fall in consumer prices was only 20 per cent in both Britain and the United States. By contrast, the range of fluctuations in industrial production increased. Between 1870 and 1913 the difference between peaks and troughs was 10 per cent in Britain and 17 per cent in the United States. Between 1920 and 1938, however, the differences were 32 per cent in Britain and 45 per cent in the United States.

The average annual growth in output per head of population in the industrialized countries between 1950 and 1973 rose to an historically high level of 3.8 per cent (and to 4.5 per cent per man-hour). The range of fluctuations fell to less than half that during the period 1890–1913, and the annual average level of unemployment came down to less than 3 per cent. But whereas between 1890 and 1913 the average annual increase in consumer prices was less than 1 per cent, between 1950 and 1973 the increase averaged over 4 per cent.

Between 1973 and 1982 the average annual growth in output per head of population fell to one-half the rate between 1950 and 1973, the range of fluctuations in industrial output doubled, and percentage unemployment trebled. Previously, apart from some brief periods at the end of major wars, such declines in output and employment led to falling prices. But between 1973 and 1982 the average annual increase in consumer prices rose to twice the 1950–73 rate.

Economists' responses

At the onset of the Great Depression of the nineteen-thirties most economists took the view that there was a strong underlying tendency for the price system to co-ordinate supplies and demands for goods and services and factors of production. Say's Law, that supply creates its own demand, had been queried from time to time since Malthus but never convincingly refuted. Say's Law did not carry the implication that in competitive market conditions there would always be a state of equilibrium. Frequent exogenous disturbances—such as wars, the discovery or exhaustion of raw materials, inventions which created opportunities for innovation and changed the relations between prices and costs—would create imbalances between supplies and demands, and it would take some time for the responses of consumers and producers to price and profit signals to restore equilibrium.

There could also be imbalances caused by the restrictive practices of monopolies, employers' associations and trade unions, or by defects in the monetary system. Thus, increases in the money supply at a greater rate than the increase in the supply of goods would raise price levels, reduce the real rate of interest below its equilibrium level and encourage levels of capital expenditure that could only be maintained by further inflation. Reductions in the rate of increase in the money supply—or even a contraction in response to previous over-expansion—would then increase the real rate of interest below its equilibrium level and bring a reduction—possibly, and in fact frequently, an abrupt reduction—in levels of capital expenditure.

The main explanation of the depressed state of the British economy in the nineteen-twenties was that patterns of production, which had been changed during the World War when international trade was restricted, were consolidated after the war by high tariffs and other restrictions on trade. The consequential reduction in the demand for the products of Britain's staple industries—iron and steel, coal, engineering, shipbuilding and textiles—created serious 'structural' unemployment, and the time taken to recover was then extended, according to Keynes and his followers, by monetary mismanagement, including the return to the gold standard on terms which overvalued sterling.

Explanations of the world economic crisis and the marked increase in unemployment between 1929 and 1932 included monetary explanations. The forecasts of Marx that capitalist crises would increase in intensity as the growth of monopolies made capitalism increasingly incapable of managing the forces of production, convinced a considerable number of economists that the recovery of employment and growth would require major changes in the economic system. Schumpeter, however, treated the Depression as just another depression following a swarm of innovations in the chemical, electrical and motor car industries, during which the new products and processes would be consolidated and the outmoded products and processes eliminated.

Keynes, who as Editor of *The Economic Journal* had published Schumpeter's paper on 'The Instability of Capitalism' in 1928, wrote in his *Treatise on Money* which was published in 1930 that it was easy to understand why fluctuations should occur in the rate of investment in fixed capital.

> Entrepreneurs are induced to embark on the production of fixed capital or deterred from doing so by their expectations of the profits to be made. Apart from the many minor reasons why these should fluctuate in a changing world, Professor Schumpeter's explanation of the major movements may be unreservedly accepted.[19]

In an essay on 'Economic Possibilities for our Grandchildren', also published in 1930, Keynes wrote of the increase in technological unemployment due to the discovery of means of economizing the use of labour faster than the discovery of new uses for labour.[20] He predicted that although much more would be heard of technological unemployment in the years to come it would prove to be only a temporary phase of maladjustment, and that in the absence of major wars or increases in population the revival of growth would bring the economic problem within sight of solution within a hundred years.

The nature of the debate on the causes and cures of unemployment was transformed by the publication of Keynes's *General Theory* in 1936. Keynes showed that when the incentive to invest was low there could be a deficiency of effective demand, and that in such a situation price and wage flexibility could not bring about full employment. The policy implication was that the maintenance of full employment depended on the willingness of governments to maintain sufficient levels of effective demand.

Although Keynes had foreshadowed a revival of growth, many of his influential followers took a less buoyant view of the prospects. There were frequent references to John Stuart Mill's chapter on the tendency to a stationary state. Their view was that the slowing of population growth in the industrial countries since 1913 and the absence of productive regions still to be settled would reduce opportunities for profitable investment more than advances in knowledge would create them. This view was more influential than Schumpeter's view that, as judged from the two previous Kondratieff cycles, there would be sufficient entrepreneurial activity and innovation between about 1940 and the mid-fifties to generate a revival, and that secondary innovations and diffusion would then create a prosperity phase lasting until the late sixties.

Schumpeter's analytical and historical account of how the sequential processes of invention, innovation and diffusion created cycles of forty months, eight to nine years, and about fifty-five years, was presented at some length in his *Business Cycles* (1939). It seems likely that the influence of that book on economists, who were later involved in the formulation of plans for post-war reconstruction and in the conduct of economic policy after the war, was reduced by its publication just before the Second World War and at a time when macroeconomists were still preoccupied with the analysis and policy prescriptions in Keynes's *General Theory.* Two other factors contributed to its lack of influence for the next thirty years. The first factor was the absence either of clear statistical evidence for the existence of regular short, medium and long cycles, or of a plausible theory of just why innovations of many different kinds—some deriving from exogenous and some from endogenous inventions, some presenting few problems of application outside the industry and some dependent on complementary innovations elsewhere—had generated and would continue to generate fairly regular cycles. The second factor was the development of the thesis in his *Capitalism, Socialism and Democracy* (1943) that the nature of the innovation process was being changed by the growth of organized industrial research and development in large corporations and the decay of the entrepreneurial function.

Few, if any, economists predicted twenty-five years of low rates of unemployment and very high rates of growth after the Second World War; and most, if not all, economists who predicted that full employment policies in capitalist economies would generate strong inflationary pressures were surprised by the moderate levels of inflation between 1948 and 1973 and by the increase in rates of inflation after the fall in growth rates and the rise in unemployment rates after 1973.

Although qualified by fears of cost-push inflation, confidence in the power of demand management policies to maintain very high levels of employment—substantially higher and more constant levels of employment than Keynes had envisaged—increased during the fifties and sixties. Even Christopher Freeman, who was familiar with Schumpeter's theories of innovation and had done pioneering research on the processes of innovation and diffusion, did not include a chapter on unemployment, or even a reference to it in the index, in the first edition of *The Economics of Industrial Innovation* published in 1974.[21]

But this growing confidence was soon undermined by the downturn in growth and employment. It is not at all surprising that economists are not of one mind on the causes of the downturn. Some treat the depression as a consequence of a shift of policy objectives away from full employment towards price stability, though the shift in policy objectives came some time after the downturn in growth and employment, and budget policies designed to counter the impact of the recession that followed the steep rise in oil prices in 1974 contributed to a marked increase in inflation rather than to an increase in employment. Others treat the depression as a consequence of a shift in R & D and investment activities towards cost-reducing and labour-displacing innovations, though if that were the explanation of the rise in unemployment there should be a productivity increase. In fact there has been a productivity slow-down. Another group of economists regards the current depression as a consequence of the maturing of the new technologies which provided the basis of the great wave of post-war growth, and of the effect of the diffusion process in reducing the profits created by the innovators. Some but not all in this group treat the present depression as the trough between the fourth and fifth Kondratieffs.

A Kondratieff trough?

In 1970–71 'when economic research was still projecting incessant economic growth and when science and technology policy were expected to produce the right kind and appropriate volume of new technology to nurture this industrial evolution',[22] Gerhard Mensch conceived and developed an analysis of industrial evolution which led him to predict a deep depression, and subsequently, if the 'rhythmical pattern in the interplay of stagnation and innovation will continue to evolve in the next 20 years as it has done in the last 300 years or so', a new 'rush' of innovation.[23]

The productivity slow-down and rise in unemployment, coming almost fifty years after Schumpeter's dating of the third Kondratieff depression—1925–34—stimulated a major revival of interest in the possibility that the inventions and introduction of new technologies generated powerful and wave-like supply-side forces which Keynesian demand management policies were not powerful enough to offset. In Holland van Duijn was an early contributor to this revival.[24] Freeman made major contributions in 1979 with his chapter in the OECD volume on *Structural Determinants of Employment and Unemployment*, in 1981 when he edited and contributed to a special edition of *Futures*, and in 1982, with his colleagues Clark and Soete, in *Unemployment and Technical Innovation*.[25]

Many economic historians and economists doubt the existence of long waves. The statistical series on rates of unemployment and growth rates in productivity during the nineteenth century are rather sketchy, but the available statistics do not provide convincing evidence of deep depressions in the troughs between the rise of the second and third long waves identified by Kondratieff and Schumpeter. The strongest supporting evidence comes from the movement of prices, not productivity and employment.

After Schumpeter's third long wave—based on electricity, chemicals and automobiles—there was a deep depression in which prices, employment rates and productivity increases declined. But in that third long wave the growth in productivity in the United States was higher between 1912 and 1925 in the supposed rundown to the trough of the long cycle than it was between 1898 and 1911 when the up-swing was supposed to be at its strongest.

What followers of Schumpeter regard as the up-swing of the fourth long wave—based on innovations in electronics, synthetics, drugs, plant breeding and pesticides—was a long period of high growth rates, low unemployment and rising prices, and in the subsequent downturn rates of employment and productivity increases fell substantially. However, the downturn was unlike any previous downturn. Prices have continued to rise, and in Britain the rate of increase in prices increased for some time after growth rates and employment rates had started to fall, and average annual price increases, which had been 5 per cent between 1938 and 1973, rose to 15 per cent between 1973 and 1979. In the industrial countries as a whole the annual rates of increase in prices rose from 4.1 per cent to 9.5 per cent.

Among those who support the view that there are long waves, there are considerable differences of opinion on the dating and the role of technical change. In *The World Economy, History and Prospect*, Rostow emphasizes the rates of change in the profitability of producing foodstuffs and raw materials and the relative prices of primary and industrial commodities. As a consequence of this emphasis, Rostow's dating of up-swing and down-swing differs, and at times differs substantially, from writers who concentrate on innovations in the secondary and service sectors. He treats 1951–73 as the down-swing of a fourth Kondratieff.[26]

Forrester treats technical change as a consequence and not as a cause of long waves, and he bases his theory of long waves in economic systems on the multiplier–acceleration mechanism that transmits the fluctuations in investment to the rest of the economy, and on capital–stock adjustment processes which cause alternating phases of capital hunger and capital saturation. When consumer demand exceeds the capacity of the existing capital equipment to supply it, the capital sector has to expand its own production capacity. But that requires a diversion of resources from the consumer goods sector that is generating the expansion of activity, and the growth in capacity is constrained. But the capital stock is eventually over-expanded, and then allowed to fall too far in the down-swing. 'The bootstrap structures tend to be highly destabilizing and to lengthen substantially the periods of fluctuation that might otherwise prevail.' In Forrester's model 'the bootstrap linkage around capital creates the 50-year cycle of what would otherwise be a 20-year medium cycle in capital acquisition', and he argues that over the past two hundred years the dynamics of fluctuations have changed very little because 'the fundamental causes revolve around the life of capital plant, the bootstrap structure around the capital sector, and the perception delays in human decision making'.[27]

In the model of endogenous fluctuations, major new technologies can only be incorporated during the early stages of successive growth periods. Once the up-swing has gone some way there is only room for minor improvements. However, this model is too dependent on stability over time in the life of capital plant, and it cannot accommodate changes in the nature of technical change which in the past have had a major influence on the extent of fluctuations.

In Schumpeter's theory 'innovations carry the business cycle', and the bunching of innovations is the key factor in long cycles. In reviewing Schumpeter's *Business Cycles*, Kuznets pointed out that to establish such a theory of long cycles it would be necessary to prove and explain the existence of discontinuities in the introduction of major innovations with a long time span.[28] Mensch attempted to do that. Although he maintained Schumpeter's four-stage long cycles, he abandoned the notion that the economy had developed in waves in favour of a theory that it has evolved through a series of innovative impulses that take the form of successive S-shaped curves.[29] In periods of stagnation and depression observed in periods around 1825, 1873 and 1928,

> the tempo of basic innovations shifts into higher gear as a new thrust of basic innovations occurs and new ideas are accommodated for the first time in the creation of novel industrial branches. . . . In the radical years 1825, 1886 and 1935 the relative frequency of innovations and the acceleration of the innovative tempo are both at their peaks. . . . If the present tendency continues, that is, delays in the transfer of knowledge followed by acceleration because of the pressure from the dammed-up demand for technological basic innovations, . . . only a small number of basic innovations that will be implemented . . . by the year 2000 will be implemented during the 1970s . . . [and] approximately two-thirds of the

> technological basic innovations that will be produced in the second half of the twentieth century will occur in the decade around 1989'.[30]

Thus, clustering is a consequence of varying degrees of pressure to introduce innovations, the pressures being strongest during depression. During the up-swing of the long wave the new major technologies become perfected and fully used, and then pseudo-innovations proliferate. During periods of prosperity, 'due to power, inertia, and other decisive reasons, investors tend to favour pseudo-innovations over basic innovations', and so the boom peters out and stagnation (stalemate) follows.

> Significant shifts in the value system lie at the root of every technological stalemate . . . the Kondratieff cycle lasts about two generations . . . stagnation reduces the usefulness and profitability of labour and capital investments in overgrown traditional business fields and thereby induces the implementation of cost-saving and product-adding innovations . . . the opposition to change must eventually give way, basic innovations are introduced that offer chances for novel investment and qualitative growth in new product and service sectors.[31]

Freeman, Clark and Soete analysed the empirical basis for Mensch's account of the innovations which followed the deep depression of the nineteen-thirties, and showed that there was a substantial and continuing flow of major innovations during the fifties and sixties, not simply a concentration of them during the thirties, and that the evidence did not support the acceleration hypothesis of reduced lead times for innovations launched in deep depressions. Mensch, they wrote, looked at the wrong swarms. Mensch also tried to prove too much. His hypothesis that major innovations are made during the depression phase is inherently implausible and not needed to explain the rise of a new wave. All that is required—apart from reductions in excess capacity and cost-push inflation—is sufficient product innovation to lay the foundations for confidence in a recovery process which would then encourage further investment in innovation.

To explain discontinuities in the innovation process, Freeman and his associates placed much greater emphasis on 'bunching' during the diffusion process, and then identified three other significant factors: first, the tendency for any pattern of innovation to shift over time from relatively labour-intensive product innovations to labour-saving process innovations; second, the tendency of the rate of profit to fall owing to the growth of capacity, the saturation of markets, and the growth of capital intensity; and third, the increase in the bargaining strength of labour following long periods of full employment. But their account of the bandwagon effect and the recession and depression phases is much more satisfactory than their account of the generation of new waves. Indeed, their account of conditions that encourage or discourage innovations implies that recovery requires exogenous stimuli. They expect the depth of depression to inhibit or delay basic inventions, and

> because of the diminishing flexibility of market economies generally, and the important role of the public sector in R & D and innovation policies, the role of public policy is crucial. But some exogenous stimulus to the system, which helps to restore the general level of profitability and improve expectations generally, may also be necessary. [p. 81.][32]

Unsettled questions

The evidence for the view that there have been four long waves generated by clusters of innovations is far from conclusive. The view would seem more plausible if there were a reasonably satisfactory theory of bunching in the innovation process. But there is not. Kuznets exposed the weakness of Schumpeter's explanation, and Freeman, Clark and Soete have shown that Mensch's theory—that the major innovations are made during the depression phase because the pressures and incentives are then at their strongest and the impediments at their weakest—does not fit the facts.

Given the extent of research and development activities, it would be surprising if there were not a revival of growth in the next few years, though whether such a revival would turn into a wave similar to the one that followed the depression of the eighteen-seventies, or to the greater wave that followed the Great Depression of the nineteen-thirties, is not an issue for which either theory or past statistical uniformities provide a good basis for forecasting.

Mensch explains the slump as a consequence of a temporary lack of useful innovations, and Freeman, Clark and Soete include in their policy prescriptions for recovery public support for measures which reduce the financial risks and gestation periods of radical new innovations. The OECD report 'Technical Change and Economic Policy' refers to such a policy as support for fundamental technologies which would provide a range of possible applications in numerous essential economic sectors.[33] Whether the flow of innovations could be made significantly more even by such measures is at this stage a matter for speculation. There is certainly a strong case for further analysis and controlled experiments.

The great range of labour-saving innovations made possible by microelectronics has persuaded some economists, including Christopher Freeman, to take seriously the possibility of job-displacing technical change.[34] Such 'job-displacing growth' would bring a secular increase in technological unemployment for the first time. During periods of mass unemployment the possibility that there has been a job-displacing tilt to technical change should be considered, though evidence of such a tilt would present difficult problems of interpretation. Freeman, Clark and Soete outlined the reasons for expecting an increase in the relative importance of labour-saving process innovations as the long wave of growth moves towards its crest and then during the recession. But there are also reasons for expecting an increase in the relative importance of labour-absorbing product innovations as a new wave of growth is generated. There is a need for much more work on the economic factors which influence changes over time in the relative strengths of job-displacing and job-creating technical changes.

The absence of any trend increase in technological unemployment is due in part to corrective forces within the economic system whenever technical change tilts further towards labour-displacement, and in part to reductions in hours of work. In Britain since 1901 the hours of work per year of full-time male workers have fallen by one-third. There is a fairly high correlation between reductions in hours and reduction in hours worked per unit output—hours fell by about one-quarter of the rate of fall in hours worked per unit output.[35] There have, however, been short periods when that relationship did not hold at all. Between 1919 and 1921 there was a very sharp reduction in hours relative to reductions in hours per unit output, and between 1948 and 1955 in increase in hours worked despite reductions in hours per unit output.

In *A Theory of Wages*, P. H. Douglas treated the reduction of hours as a result of workers' decisions to reduce hours of work when, as a consequence of productivity increases, hourly rates of pay increased.[36] Such decisions to take part of the increased productivity as additional leisure rather than as additional income reduced the capacity to produce and probably the proportion of income saved below what they would otherwise have been. The supply of labour was also reduced by a rise in the average age of entry to the labour market and a fall in the average age of retirement. That reduction in the average number of years spent in the labour force was brought about by the effects of higher incomes on family expenditure patterns, and by government decisions, as influenced by the rise in tax capacity, to increase expenditures on education and pensions.

Since the mid-seventies there has been a growing tendency to argue that there is a need for greater attention to changes in hours of work and years in the labour-force as components in employment policy. The results of econometric studies of the effects of reducing hours depend on the assumptions about the extent to which hourly rates are increased when hours are reduced and on whether 'Keynesian' or 'monetarist' models are used. Just what type of econometric model should be used to test such policy issues is an important unsettled question to which perhaps the next generation of economists will provide the answer. The passage of time will provide the answer to the other unsettled question—whether there is more than a temporary shortage of labour-absorbing innovations.

Notes

1. A. Maddison, *Phases of Capitalist Development*, Oxford and New York, Oxford University Press, 1982, pp. 35 and 115–23.
2. Adam Smith, *An Enquiry into the Nature and Causes of the Wealth of Nations* (1776), Dent edition 1910, Book 1, Chap. 1.
3. Quoted in Keynes, *Essays in Persuasion*, (London: Macmillan, 1933), p. 142.
4. *Principles of Political Economy*, (London: Everymans Library Edition, 1911), p. 264.
5. *Principles of Political Economy*, (London: People's Edition, 1865), Bk. 4, Chap. 4.
6. Ibid., Bk. 4, Chap. 2.
7. K. Marx and F. Engels, *The Communist Manifesto*, London, Penguin edition, 1967, pp. 81, 83.
8. K. Marx, *Grundrisse*, London, Penguin edition, 1973, pp. 703–4.
9. W. Rostow, *British Economy in the Eighteenth Century*, Oxford, Clarendon Press, 1948, pp. 45, 48.
10. A. Maddison, *Phases of Capitalist Development*, Oxford, Oxford University Press, 1982, Chap. 4 and Appendix E.
11. N. D. Kondratieff, 'Die langer Wellen der Konjunktur', *Archiv für Sozialwissenschaft und Sozialpolitik*, Vol. 56 (1926), pp. 573–609. English translation by W. F. Stolper, 'Long Waves in Economic Life', *Review of Economic Statistics*, **17**, No. 6 (November 1935), pp. 105–15.
12. S. Kuznets, *Secular Movements in Movements in Prices and Production*, Boston, Houghton Mifflin, 1930.
13. J. A. Schumpeter, *Business Cycles*, New York, McGraw Hill, 1939.
14. J. A. Schumpeter, *Theorie der Wirtschaftlichen Entwicklung*, Leipzig: Duncker and Humboldt, 1912. English translation, Harvard University Press, 1934.
15. Maddison, op. cit., Appendix C.
16. J. A. Schumpeter, *Capitalism, Socialism and Democracy*, London, Allen and Unwin, 1943, p. 64.
17. For the United States, Yale Brozen estimated that R & D expenditure in manufacturing industry grew from less than 0.5 per cent of sales in the thirties and less than 1 per cent in the forties, to 1.7 per cent in the fifties. See 'The Future of Industrial Research' in *The Rate and Direction of Inventive Activity*, Princeton, Princeton University Press, 1962, pp. 273–6.
18. J. A. Schumpeter, 'The Instability of Capitalism', *The Economic Journal*, **38**, No. 151 (1928), pp. 361–86. Fifteen years later, in *Capitalism, Socialism and Democracy*, New York, Harper and Row, 1943, he predicted that through organized R & D large corporations would produce innovations to order.
19. J. M. Keynes, *Treatise on Money*, London, Macmillan, 1930, Vol. 2, p. 85.
20. Included in J. M. Keynes, *Essays in Persuasion*, London, Macmillan, 1933.

21. C. Freeman, *The Economics of Industrial Innovation*, London, Penguin Modern Economics Texts, 1974. The second edition, published in London by Frances Pinter in 1982, was extended to include a chapter on 'Unemployment and Policies for Technology'.
22. G. Mensch, *Stalemate in Technology*, Ballinger, Cambridge, Mass., 1979, pp. xvii–xviii and 189. German edition, *Das technologische Patt*, Frankfurt, Umschau Verlag, 1985.
23. Ibid., p. 189.
24. A convenient reference is his paper 'Fluctuations in Innovation over Time', *Futures*, **13** (1981), pp. 264–75.
25. The OECD volume was published in Paris in 1979, and the paper in *Futures* **13** was called 'Technical Innovation and Long Waves in Economic Development'. *Unemployment and Technical Innovation: A Study of Long Waves and Economic Development* was published by Frances Pinter, London, in 1982.
26. W. Rostow, *The World Economy, History and Prospect*, London, Macmillan, 1978.
27. J. W. Forrester,'Growth Cycles' in *The Economist*, **125** (1977), pp. 534, 536. See also his chapter on 'Innovation and Social Change' in *Long Waves in the World Economy*, C. Freeman (ed.), London, Frances Pinter, 1984, pp. 126–34.
28. S. Kuznets, 'Schumpeter's Business Cycles', in *American Economic Review*, **30** (1940), pp. 257–71.
29. Mensch, op. cit., note 22, p. 73.
30. Ibid., pp. 175, 177, 197.
31. Ibid., pp. xvii, 5–7.
32. C. Freeman, J. A. Clark and L. G. Soete, op. cit. (note 25), p. 81. In his 'Background Material for a Meeting on Long Waves, Depression and Innovation', IIASA, August 1983, Freeman referred to the need for a new international framework and added the comment that the development of such a 'framework for expansion may prove to be the most difficult problem confronting the world economy if there is to be a fifth Kondratieff upswing'.
33. OECD, *Technical Change and Economic Policy*, Paris, 1980, pp. 101–2. Professor Freeman was a member of the group that produced the report.
34. Sol Encel and Jarlath Ronayne (eds), *Science, Technology and Public Policy*, Pergamon Press, Sydney, Australia, 1979, pp. 53–76.
35. P. Armstrong, *Technical Change and Reductions in Life Hours of Work*, London, The Technical Change Centre, 1984.
36. P. H. Douglas, *A Theory of Wages*, New York, Macmillan, 1934.

PART IV

Viewing the Human Prospect

9 Schumpeter and Marx: how common a vision?*

Nathan Rosenberg

> 'Whereas a stationary feudal economy would still be a feudal economy, and a stationary socialist economy would still be a socialist economy, stationary capitalism is a contradiction in terms.'[1]

I

Recent years have witnessed a resurgence of interest in Schumpeter's work. The reasons are not far to seek. The increasing realization of the major role of technological change in generating economic growth has been a powerful stimulus, in view of Schumpeter's central preoccupation with the innovation process. The search for the determinants of innovations has led to a reexamination of Schumpeterian notions of the relationship between industrial structure and innovative performance. Thus, there have been attempts to formulate more rigorous theories of Schumpeterian competition, and to marshall empirical data that would illuminate more clearly the relationship between size of firm, industrial concentration, and innovative performance. Moreover, the recent heightened interest in the possibility of long waves has also led to a reexamination of Schumpeter's careful treatment of the subject, a subject to which Chris Freeman has made notable contributions.[2]

I believe that Schumpeter was one of the most profoundly interesting economists of the twentieth (or, for that matter, of any other) century. I therefore welcome this Schumpeterian renaissance as a healthy intellectual development in modern economics. However, I want to argue that this renaissance has, so far, been an excessively partial one. That is, it has confined itself to one rather restricted portion of a much larger body of thought. One of Schumpeter's greatest strengths was that he took a very wide-ranging view of the economic process. He saw this process as a part of a larger social and historical frame of reference. He possessed, moreover, a complex vision of the economic process—and I use the term 'vision' in Schumpeter's own precise sense, which I will delineate shortly.

* I have incurred substantial obligations to Stanley Engerman, Karl Habermeier and Edward Steinmueller for their comments on earlier versions of this paper.

That vision had powerful implications for the usefulness of much of the neo-classical economics that was dominant in Schumpeter's time. In fact, it bears certain very intriguing connections with Marx's earlier vision of the economic process.

Thus, it strikes me that there may be something to gain from an examination of the Schumpeterian and Marxian visions. In undertaking this examination, I will focus primarily upon Schumpeter's writings and allude to Marx more briefly. The justification is that Marx's views are by now widely known, whereas many of Schumpeter's contributions to economic and social thought remain neglected—even by people who would not shrink from the label 'Neo-Schumpeterians'.

Schumpeter often proclaimed himself an admirer of Marx (but not of 'vulgar Marxism'), and he wrote extensively and perceptively about him.[3] I wish to suggest that Schumpeter and Marx shared a common vision, some elements of which Schumpeter was anxious to acknowledge, but others which he did not, and perhaps would even have denied.

II

By 'vision' Schumpeter referred to a set of phenomena that one perceives as being related in some significant ways. That perception, and the judgment that the phenomena in question are significant and worthy of scientific analysis, is something that necessarily precedes and forms the basis of scientific research. It is not science but, rather, a precondition for the conduct of science:

> in order to be able to posit to ourselves any problems at all, we . . . first have to visualize a distinct set of coherent phenomena as a worth-while object of our analytic efforts. In other words, analytic effort is of necessity preceded by a preanalytic cognitive act that supplies the raw material for the analytic effort. In this book, this preanalytic cognitive act will be called Vision.[4]

Thus, one's vision serves as a kind of selective mechanism for identifying those aspects and interrelations in the objective world that one deems significant and therefore worthy of analytic and empirical effort.

In this sense, Marx and Schumpeter held in common a vision of capitalism as a social system that possessed its own internal logic, and that underwent a process of self-transformation. This self-transformation resulted from certain 'laws of motion,' as Marx called them, which were inherent in capitalism as a social system. Thus, it was possible to understand the dynamics of capitalism as the system actually behaves over historical time if one could grasp those laws of motion. In other words,

Schumpeter believed that it was possible to develop a theory of economic change. This is in stark contrast to the static equilibrium model prevailing in economic theory at that time, which examined how the economy re-established itself to its equilibrium position after a small disturbance. In his preface to the Japanese edition of *The Theory of Economic Development* (1937), Schumpeter observed:

> If my Japanese readers asked me before opening the book what it is that I was aiming at when I wrote it, more than a quarter of a century ago, I would answer that I was trying to construct a theoretic model of the process of economic change in time, or perhaps more clearly, to answer the question how the economic system generates the force which incessantly transforms it. . . . I felt very strongly . . . that there was a source of energy within the economic system which would of itself disrupt any equilibrium that might be attained. If this is so, then there must be a purely economic theory of economic change which does not merely rely on external factors propelling the economic system from one equilibrium to another. . . .[5]

This statement of intellectual purpose would also have served as an apt prolegomenon to Marx's work.[6] Marx and Schumpeter possessed strikingly similar intellectual agendas—which is merely another way of stating that they shared a common vision. Several of Marx's primary concerns are also the primary concerns of Schumpeter's analysis of capitalism:

1. The growth of the size of the firm and in industrial concentration (the central theme of *Capitalism, Socialism and Democracy*);
2. the inherent instability of capitalism and the inevitability of 'crises' (Schumpeter's major concern, as embodied in his most ambitious work, *Business Cycles*); and
3. the eventual destruction of capitalist institutions, and the arrival of a socialist form of economic organization, as a result of the working out of the internal logic of capitalist evolution (a theme intimately connected with the growth of industrial bigness, spelled out in detail in *Capitalism, Socialism and Democracy*).

I do not wish to ruin what I regard as a good argument by overstatement. The reader is of course familiar with fundamental differences between these two seminal figures. The 'details' of the process of economic change visualized by Marx, and leading to socialism, involve 'immizerization' of the proletariat; whereas, for Schumpeter, it is precisely the highly successful economic performance of capitalism that generates political and social forces leading to 'Crumbling Walls' (the title of Chapter 12 of *Capitalism, Socialism and Democracy*) and eventual breakdown.[7] At the same time, Schumpeter himself was anxious to point out that, in certain respects at least, it is easy to exaggerate the *differences* between his own analysis and that of Marx:

> The capitalist process not only destroys its own institutional framework but it also creates the conditions for another. Destruction may not be the right word after all. Perhaps I should have spoken of transformation. The outcome of the process is not simply a void that could be filled by whatever might happen to turn up; things and souls are transformed in such a way as to become increasingly amenable to the socialist form of life. With every peg from under the capitalist structure vanishes an impossibility of the socialist plan. In both these respects Marx's *vision* was right. . . . In the end there is not so much difference as one might think between saying that the decay of capitalism is due to its success and saying that it is due to its failure.[8]

III

I now propose to examine some of the separate components of Schumpeter's vision in more detail, still against the backdrop of Marx. First, capitalism has to be understood as an evolutionary system rather than as a system that continually reverts to equilibrium after small departures from it, as most bourgeois economists of Schumpeter's day viewed it. More specifically, this evolution is a reflection of certain dynamic forces that Schumpeter, along with Marx, believes are inherent in the incentive structure, the drive toward profits, and the competitive institutions of the system. A system of analysis that abstracts from such change is a system that abstracts from the very essence of capitalism. On this point, Schumpeter is anxious to associate himself with Marx and to remind the reader that the theories of perfect and imperfect competition systematically neglect this central feature.[9]

> Capitalism . . . is by nature a form or method of economic change and not only never is but never can be stationary. And this evolutionary character of the capitalist process is not merely due to the fact that economic life goes on in a social and natural environment which changes and by its change alters the data of economic action; this fact is important and these changes (wars, revolutions and so on) often condition industrial change, but they are not its prime movers. Nor is this evolutionary character due to a quasi-automatic increase in population and capital or to the vagaries of monetary systems of which exactly the same thing holds true.[10]

The implications of this view are, for Schumpeter, very far-reaching. For, as he goes on to argue, it renders essentially irrelevant that view of the competitive process which concerns itself with 'how capitalism administers existing structures' since 'the relevant problem is how it creates and destroys them.' As soon as this is recognized, one's 'outlook on capitalist practice and its social results changes considerably.'

> The first thing to go is the traditional conception of the *modus operandi* of competition. Economists are at long last emerging from the state in which price competition was all they saw. As soon as quality competition and sales effort are admitted into the sacred precincts of theory, the price variable is ousted from its dominant position. However, it is still competition within a rigid pattern of invariant conditions, methods of production and forms of industrial organization in particular, that practically monopolizes attention. But in capitalist reality as distinguished from its textbook picture, it is not that kind of competition which counts but the competition from the new commodity, the new technology, the new source of supply, the new type of organization (the largest-scale unit of control for instance)—competition which commands a decisive cost or quality advantage and which strikes not at the margins of the profits and the outputs of the existing firms but at their foundations and their very lives.[11]

Thus, the main agency making for economic change is innovation, a phenomenon that is defined much more broadly, as the last quotation makes clear, than only technological innovation. Nevertheless, technological innovation is central to long-term economic change for Schumpeter, just as it had been for Marx. Moreover, Schumpeter and Marx share a common view on the determinants of science and technology that sets them distinctly apart from the main line of thinking in economics, where these have been treated as exogenous variables. Specifically, they regard the remarkable performances of *both* science and technology in the western world as having been overwhelmingly due to capitalism and the associated bourgeois culture.

Schumpeter asserts in his later work[12] that progress in both science and technology must be understood to have been endogenous to capitalism, and he insists that this was Marx's view as well. Schumpeter observes that Marx had, in the *Communist Manifesto*, 'launched out on a panegyric upon bourgeois achievement that has no equal in economic literature.' After quoting a relevant portion of the text,[13] he says:

> No reputable 'bourgeois' economist of that or any other time—certainly not A. Smith or J. S. Mill—ever said as much as this. Observe, in particular, the emphasis upon the *creative* role of the business class that the majority of the most 'bourgeois' economists so persistently overlooked and of the business class *as such*, whereas most of us would, on

> the one hand, also insert into the picture non-bourgeois contributions to the bourgeois success—the contributions of non-bourgeois bureaucracies, for instance—and, on the other hand, commit the mistake (for such I believe it is) to list as independent factors science and technology, whereas Marx's sociology enabled him to see that these as well as 'progress' in such fields as education and hygiene were just as much the products of the bourgeois class culture—hence, ultimately, of the business class—as was the business performance itself.[14]

IV

These statements would not be completely surprising if held by someone who, with Marx, believes in the economic interpretation of history. But does that include Schumpeter? I believe it does, with certain qualifications, but the qualifications that he imposed upon the economic interpretation of history are of a sort that, if anything, actually strengthen its usefulness as a device for explaining economic change. Thus, I argue that, here, too, Schumpeter and Marx shared a common vision.

Schumpeter in several places offers a sympathetic and approving treatment of the economic interpretation of history; moreover, almost all of his writing fits conveniently into that interpretation. He defends Marx against some of the cruder criticisms of the economic interpretation, such as the view that it reduces all human behaviour to narrowly-based economic motives, or that economic *materialism* was somehow logically incompatible with metaphysical or religious beliefs.[15]

Schumpeter reduces the economic interpretation to two propositions:

1. The forms or conditions of production are the fundamental determinants of social structures which in turn breed attitudes, actions and civilizations.
2. The forms of production themselves have a logic of their own; that is to say, they change according to necessities inherent in them so as to produce their successors merely by their own working.[16]

Schumpeter asserts that 'Both propositions undoubtedly contain a large amount of truth and are, as we shall find at several turns of our way, invaluable working hypotheses.'[17] His main 'qualification,' if that is what it really is, is his insistence upon the importance of lags, i.e. social forms that persist after they have lost their economic rationale. It is far from clear that Marx would have disagreed with such a qualification, since Marx was much too sophisticated a historian to believe that economic changes generated the 'appropriate' social changes instantaneously. Schumpeter, in making the qualification about lags, adds that Marx, although perhaps not fully appreciating their implications, would not have taken the simplistic position involved in denying them a role.

> Social structures, types and attitudes are coins that do not readily melt. Once they are formed they persist, possibly for centuries, and since different structures and types display different degrees of this ability to survive, we almost always find that actual group and national behaviour more or less departs from what we should expect it to be if we tried to infer it from the dominant forms of the productive process. Though this applies quite generally, it is most clearly seen when a highly durable structure transfers itself bodily from one country to another. The social situation created in Sicily by the Norman conquest will illustrate my meaning. Such facts Marx did not overlook but he hardly realized all their implications.[18]

V

The final points that I wish to deal with have really been implicit in some of what has already been said. Schumpeter, like Marx, was attempting to develop an analytical apparatus that could be employed in accounting for economic change. In pursuing this goal, Schumpeter mounted a set of criticisms of neo-classical economics that deserve to be regarded as among the most devastating that have ever been formulated. Among the most important objections Schumpeter raised were his criticisms of the concepts of rationality and consumer sovereignty, both of which spring from his fundamentally different conception of the nature of innovation and the implications this has for the treatment of economic behaviour under uncertainty.

Of course it should be pointed out that the neo-classical economics of Schumpeter's time has been largely superseded by what may be called, for convenience, 'modern economics,' a body of analysis that incorporates many of the tenets and techniques of neo-classical economics, but goes far beyond it in numerous ways. On the surface, at least, modern economics appears to have addressed many of the problems that Schumpeter saw in neo-classical economics. Intertemporal and temporary general equilibrium theory have been developed, the use of dynamic game theory is becoming widespread, models of asymmetric information are applied to questions of institutional functioning, and social choice theory seeks to tackle many of the thorny issues of political economy. Changes in taste have been investigated and decision theorists have debated the question of rationality. We even have a neo-classical theory of economic growth. Nevertheless, it is my own view that, at a deeper level, much of what Schumpeter had to say about neo-classical economics applies to modern economics as well.

It is not a sufficient response to say that Schumpeter was simply interested in a different set of problems than are economists in the neo-classical and modern traditions, although that is of course both true and important. In Schumpeter's view, there is a dichotomy between 'vision' and 'economic analysis.' Given Schumpeter's vision of the economic self-transformation of the social system, the function of refinements in analysis is to develop or to disprove specific empirical statements. Rather, what is probably the most important difference between Schumpeterian and neo-classical/modern economics is the absence in neo-classical and modern economics of a 'vision' of capitalism as an essentially historical phenomenon and subject to a historical process of transformation. Neo-classical and, to an even greater extent, modern theory, operate in a historical vacuum, attempting to develop principles of human behaviour and economic organization that are valid without reference to time or place, and most often even independent of social organization. Historical counter-examples are usually 'integrated' by making the concepts and relationships of the theory still more general; for which of course a high price has to be paid in terms of its ability to predict specific empirical phenomena. This leads to an incoherent multiplicity of *ad hoc* specifications in individual applications. An overall view of economy and society as a historical process is lacking. In fact, there are some modern theorists who would regard such an overall view as meaningless.

As I have attempted to show earlier, Schumpeter believed that innovation is the central feature of the capitalist system. The process of innovation cannot be reduced to mere calculation, although imitation of the innovation can. Innovation is the creation of knowledge that cannot, and therefore should not, be 'anticipated' by the theorist in a purely formal manner, as is done in the theory of decision-making under uncertainty. In Schumpeter's view, it would be entirely meaningless to speak of 'the future state of the world,' as that state is not merely unknown, but also indefinable in empirical and historical terms. Serious doubt is thus cast on what meaning, if any, can be possessed by inter-temporal models of equilibrium under uncertainty, in which the essential nature of innovation is systematically neglected.

> The assumption that business behaviour is ideally rational and prompt, and also that in principle it is the same with all firms, works tolerably well only within the precincts of tried experience and familiar motive. It breaks down as soon as we leave those precincts and allow the business community under study to be faced by—not simply new situations, which also occur as soon as external factors unexpectedly intrude but by—new possibilities of business action which are as yet untried and about which the most complete command of routine teaches nothing.

> Those differences in the behaviour of different people which within those precincts account for secondary phenomena only, become essential in the sense that they now account for the outstanding features of reality and that a picture drawn on the Walras-Marshallian lines ceases to be true—even in the qualified sense in which it is true of stationary and growing processes: it misses those features, and becomes wrong in the endeavour to account by means of its own analysis for phenomena which the assumptions of that analysis exclude.[19]

In this respect, then, Schumpeter's vision of the essential nature of capitalism is one that causes him to relegate neo-classical analysis to a very subordinate position, that of the analysis of secondary phenomena, if not mere epiphenomena. That is to say, neo-classical analysis is useful in helping to understand how capitalist reality would perform when that system is deprived of its most distinctive features, including the historical context without which those features cannot be defined.

This has serious implications for the concept of rationality, and thus for welfare economics and for the meaning of models of decision making under uncertainty. If rationality is reduced in neo-classical and modern theory more and more to the tautology that people do the best they can, given the whole gamut of constraints they face, among the most important of which is the informational constraint, then accepting Schumpeter's concept of innovation means that human actions are always second-best in a way which ultimately cannot be subjected to further analysis. Rational behaviour, then, is most significant in a world of routine and repetition of similar events.[20]

This is a complex issue, for, according to Schumpeter, capitalism needs to be perceived as a historical force that encourages rational calculation. Historically, some of the most significant accomplishments of capitalism were connected with the ways in which the system progressively expanded the social sphere within which rational calculation was possible. This is precisely the theme of Chapter 11 of *Capitalism, Socialism and Democracy*, 'The Civilization of Capitalism.'[21] But for Schumpeter this is perfectly compatible with the view that the central dynamic of capitalism—innovation—is a social activity that goes far beyond mere rational calculation.[22]

Along a slightly different line, Schumpeter attacked the fundamental notion that preferences were exogenously given.

> Innovations in the economic system do not as a rule take place in such a way that first new wants arise spontaneously in consumers and then the productive apparatus swings round through their pressure. We do not deny the presence of this nexus. It is, however, the producer who as a rule initiates economic change, and consumers are educated by him if

necessary; they are, as it were, taught to want new things, or things which differ in some respect or other from those which they have been in the habit of using. Therefore, while it is permissible and even necessary to consider consumers' wants as an independent and indeed the fundamental force in a theory of the circular flow, we must take a different attitude as soon as we analyse *change*.[23]

The same essential point was made later on, in *Business Cycles*:

> We will, throughout, act on the assumption that consumers' initiative in changing their tastes—i.e., in changing that set of our data which general theory comprises in the concepts of 'utility functions' or 'indifference varieties'—is negligible and that all change in consumers' tastes is incident to, and brought about by, producers' actions. This requires both justification and qualification.
>
> The fact on which we stand is, of course, common knowledge. Railroads have not emerged because any consumers took the initiative in displaying an effective demand for their service in preference to the services of mail coaches. Nor did the consumers display any such initiative wish to have electric lamps or rayon stockings, or to travel by motorcar or airplane, or to listen to radios, or to chew gum. There is obviously no lack of realism in the proposition that the great majority of changes in commodities consumed has been forced by producers on consumers who, more often than not, have resisted the change and have had to be educated up by elaborate psychotechnics of advertising.[24]

Although modern economists have investigated the consequences of endogenous preferences for welfare judgments, most have considered it better, for reasons of division of labor with other disciplines, in particular psychology, to neglect the investigation of why and how tastes change.[25] Schumpeter asserted that innovation, the fundamental driving force of the historical evolution of capitalism, would mould tastes as well as technology in unexpected ways. The implications, both for the development of the economic and social systems, as well as for microeconomic welfare judgments, were potentially radical. Just before his death in 1950 he severely criticized economists for 'the uncritical belief that so many seem to harbour in the virtues of consumer's choice.'

> First of all, whether we like it or not, we are witnessing a momentous experiment in malleability of tastes—is not this worth analyzing? Second, ever since the physiocrats (and before), economists have professed unbounded respect for the consumers' choice—is it not time to investigate what the bases for this respect are and how far the traditional and, in part, advertisement-shaped tastes of people are subject to

the qualification that they might prefer other things than those which they want at present as soon as they have acquired familiarity with these other things? In matters of education, health, and housing there is already practical unanimity about this—but might the principle not be carried much further? Third, economic theory accepts existing tastes as data, no matter whether it postulates utility functions or indifference varieties or simply preference directions, and these data are made the starting point of price theory. Hence, they must be considered as independent of prices. But considerable and persistent changes in prices obviously do react upon tastes. *What, then, is to become of our theory and the whole of micro-economics?* It is investigations of this kind, that might break new ground, which I miss.[26]

Thus, Schumpeter's explicit criticisms of neo-classical economics, and his implicit criticisms of the modern economics which grew from it, are devastating. He is, after all, led by the logic and by the implications of his own theory to express the greatest skepticism concerning the validity of some of the most central components of economic theory.[27] These criticisms even include the beneficent outcomes that economists have consistently attributed to the competitive process. If one accepts the Schumpeterian view that 'capitalist reality is first and last a process of change,'[28] then that reality feeds back upon the functioning of competition as well. For, as Schumpeter states: 'In appraising the performance of competitive enterprise, the question whether it would or would not tend to maximize production in a perfectly equilibrated stationary condition of the economic process is hence almost, though not quite, irrelevant.'[29]

The reason it is not completely irrelevant is that the model of a stationary competitive process helps to understand the behaviour of an economy that possessed no internal forces generating economic change. Thus, the model of a Walrasian circular flow constitutes Schumpeter's starting-point in understanding the essential elements of capitalist reality because it shows how the system would behave in the absence of its most distinctive feature—innovation.[30] It is an invaluable abstraction precisely because it makes it possible to trace out with greater precision the impact of innovative activity. The circular flow serves as the analytical starting-point for Schumpeter's theory of business cycles and growth, and the periodic tendency of the economic system to revert to an equilibrium is an indispensable component of that theory.[31]

It is important to understand the methodological use that Schumpeter makes of the neo-classical analysis of a stationary economic process. One cannot draw valid inferences from such analyses about the efficiency with which actual capitalistic economies allocate resources. Rather, if one accepts his argument, innovation is the essential feature of capitalist reality, and this particular aspect of reality dominates the manner in which the system performs.[32] Thus, the analysis of a stationary economic process is justifiable as a starting-point for the study of capitalist institutions, but not as a *terminus*.

VI

I conclude that Schumpeter and Marx had very common visions. But by this I do *not* mean just the common element that Schumpeter himself liked to emphasize—the vision of the breakdown of capitalism and the inevitability of socialism. Rather, the point that I want to stress—and it is of course related to this point—is that they both saw the essential role of economics as consisting in accounting for economic change as a historical phenomenon. And, as a result, they both emphasized the inadequacy of a methodology that concerns itself only with small incremental changes and the nature of the associated adjustment mechanisms. They were both interested in *transformation* processes, not adjustment processes.

In Schumpeter's case, though he never faced up to it, his analysis really amounted to a wholesale rejection of some of the basic tenets of neo-classical reasoning. Schumpeter believed that neo-classical analysis did not provide an adequate framework for understanding the essential aspects of capitalist reality. In spite of Schumpeter's numerous expressions of filial piety to Walras, including his description of Walras' *Elements* as 'this Magna Charta of exact economics',[33] his rejection of neo-classical reasoning is profound.

I am not certain how I should like to resolve that paradox, beyond calling attention to the obvious fact that Schumpeter was a lover of paradox. It is, after all, an equal paradox that Schumpeter, a man of deep conservative instincts, should have written far more (and far more convincingly) about the essential economic workability of a socialist society than did Marx, the foremost exponent of socialism.[34]

Perhaps Schumpeter's great sensitivity to the limitations of neo-classical reasoning was intensified by his deep immersion in the works of the German Historical School. In the German-speaking world of Schumpeter's formative years, the *Methodenstreit* between the historical and theoretical schools was a standing intellectual fact of life, and Schumpeter was of course intimately familiar with, and respectful of, the works of Gustav Schmoller, Werner Sombart and Max Weber. Under these circumstances he might be expected to be well aware of the limitations of an approach to economic life which made no serious pretence of dealing with economic behaviour in real historical time. Thus, Schumpeter was in an excellent position to recognize the possibilities and the limitations of the purely analytical and the purely historical approaches to economics. His own approach really involved a complex mixture of Marxism, Walrasian equilibrium analysis, and German historical scholarship. It was not a fusion of all three, because the three are not capable of being completely fused. But it was a brilliant mixture, if not always an internally consistent one.

Notes

1. Joseph A. Schumpeter, 'Capitalism in the Postwar World', in R. Clemence (ed.), *Essays of J. A. Schumpeter*, Cambridge, Mass., Addison-Wesley, 1951, p. 174.
2. C. Freeman *et al.*, *Unemployment and Technological Innovation:A Study of Long Waves and Economic Development*, London, Frances Pinter, 1982.
3. See, in particular, the first four chapters of *Capitalism, Socialism and Democracy*, London, Allen and Unwin, 1943; '*The Communist Manifesto* in Sociology and Economics', *Journal of Political Economy*, **57** (1949), pp. 199–212, as reprinted in *Essays*, op. cit., note 1, pp. 282–95; and the scattered discussions in *History of Economic Analysis*, New York, Oxford University Press, 1954.
4. Cf. *History*, op. cit., note 3, p. 41. For Schumpeter's treatment of the closely related issue of ideology, see his Presidential Address to the American Economic Association, 'Science and Ideology', *American Economic Review*, **39** (1949), pp. 345–59, as reprinted in *Essays*, op. cit., note 1, pp. 267–81.
5. As reprinted in *Essays*, op. cit., note 1, pp. 158–60.
6. Indeed, there are some striking parallels in Marx's preface to the German edition of *Capital*. Schumpeter says of Marx, in *Capitalism, Socialism and Democracy*, op. cit., note 3: 'Through all that is faulty or even unscientific in his analysis runs a fundamental idea that is neither—the idea of a theory, not merely of an indefinite number of disjointed individual patterns or of the logic of economic quantities in general, but of the actual sequence of those patterns or of the economic process as it goes on, under its own steam, in historic time, producing at every instant that state which will of itself determine the next one. . . . Economists always have either themselves done work in economic history or else used the historical work of others. But the facts of economic history were assigned to a separate compartment. They entered theory, if at all, merely in the role of illustrations, or possibly of verification of results. They mixed with it only mechanically. Now Marx's mixture is a chemical one; that is to say, he introduced them into the very argument that produces the results. He was the first economist of top rank to see and to teach systematically how economic theory may be turned into historical analysis and how the historical narrative may be turned into *histoire raisonée.*' (pages 43–4.)
7. It is also appropriate to note that the main interest of the economics profession has always been in Schumpeter's theory of the business cycle rather than in his theory of economic change. Some years ago, Paul Sweezy asserted: 'It is noteworthy that in orthodox circles, Schumpeter's theory of economic development has never commanded anything like the attention which it deserves and that it has been widely misunderstood and misrepresented. In so far as it has achieved recognition it has done so as business cycle theory rather than as the foundation of a theory of capitalist evolution. In the final analysis,

therefore, the example of Schumpeter serves only to emphasize the modern orthodox economist's lack of interest in what Marx called capitalism's "law of motion."' (Paul Sweezy, *Theory of Capitalist Development*, New York, Oxford University Press, 1942, p. 95.) I agree with Sweezy's statement, but I feel that it is also noteworthy how little interest Marxists have expressed in Schumpeter's work, in spite of the fact that this work addresses such a familiar range of problems and was animated by such a similar vision. One cannot suppress the suspicion that the explanation may lie in Schumpeter's very positive evaluation of the performance of late capitalism *as an economic system.*

8. Cf. *Capitalism, Socialism and Democracy*, op. cit., note 3, p. 162 (emphasis Schumpeter's). It is of interest to note that Schumpeter's basic diagnosis for the future of capitalism had been foreshadowed in an article in the *Economic Journal* as early as 1928. He concluded the paper with the following one-sentence summary: 'Capitalism, whilst economically stable, and even gaining in stability, creates, by rationalizing the human mind, a mentality and a style of life incompatible with its own fundamental conditions, motives and social institutions, and will be changed, although not by economic necessity and probably even at some sacrifice of economic welfare, into an order of things which it will be merely matter of taste and terminology to call Socialism or not.' (Joseph Schumpeter,'The Instability of Capitalism', *Economic Journal*, **38** (1928), pp. 361–86, as reprinted in *Essays*, op. cit., note 1, pp. 71–2.)
9. See Schumpeter's criticisms of the theories of perfect competition, imperfect competition and monopolistic competition in the last pages of Chapter 6 and in Chapter 7 of *Capitalism, Socialism and Democracy*, op. cit., note 3. Of the conclusion, drawn from the theory of imperfect competition, that the typical firm will be sub-optimally small, Schumpeter sardonically observes: 'Since imperfect competition is . . . held to be an outstanding characteristic of modern industry we are set to wondering what world these theorists live in, unless fringe-end cases are all they have in mind.' Ibid., p. 85, fn. 4.
10. Ibid., pp. 82–3.
11. Ibid., pp. 84–5. Since the most important form of competition comes from innovation, and large firms are more effective innovators, the welfare and efficiency arguments that are traditionally drawn concerning perfectly competitive markets are no longer tenable. Even the criticism of monopoly power as leading to socially undesirable, restrictive practices is no longer appropriate. 'Thus it is not sufficient to argue that because perfect competition is impossible under modern industrial conditions—or because it always has been impossible—the large-scale establishment or unit of control must be accepted as a necessary evil inseparable from the economic progress which it is prevented from sabotaging by the forces inherent in its productive apparatus. What we have got to accept is that it has come to be the most powerful engine of that progress and in particular of the long-run expansion of total output not only in spite of, but to a considerable extent through, this strategy which looks so restrictive when viewed in the individual case and from the individual point of time. In this respect, perfect competition is not only impossible but inferior, and has no title to being set up as a model of ideal efficiency.' (Ibid., p. 106.)
12. Schumpeter's views in one important respect changed substantially over his lifetime. In his early *Theory of Economic Development* (Cambridge, Mass., Harvard University Press, 1949, first published in 1912) and even in his later *Business Cycles* (New York, McGraw-Hill, 1939), invention is treated as

exogenous, or at least its behaviour and rate are not explicitly influenced by economic phenomena. On the other hand, the actual *adoption* of inventions is treated as not only endogenous but as a central fact of Schumpeter's business cycle theory.

By the time he wrote *Capitalism, Socialism and Democracy* (op. cit., note 3) in the late 1930s and early years of the Second World War, Schumpeter placed great emphasis on the economic determinants of invention. Or, more precisely, he argued that both invention and innovation were becoming institutionalized and 'automated' in the large corporation, a development that substantially—and successfully—bureaucratized the entrepreneurial function. One might argue that there was a simple reason for Schumpeter's change of view: the capitalist world itself changed very drastically in the first forty years or so of the twentieth century, and Schumpeter's changing view simply recorded a significant change in capitalist reality.

13. Schumpeter quotes the assertions of Marx and Engels that the bourgeoisie 'has been the first to show what man's activity can bring about. It has accomplished wonders far surpassing Egyptian pyramids, Roman aqueducts, and Gothic cathedrals' and 'by the rapid improvement of the instruments of production . . . draws all nations, even the most barbarian, into civilization . . . it has created enormous cities' and 'during its rule of scarce one hundred years has erected more massive and more colossal productive forces than have all preceding generations together.' ('*The Communist Manifesto* in Sociology and Economics', op. cit., note 3, p. 293.)
14. Ibid., p. 293. See also *Capitalism, Socialism and Democracy*, op. cit., note 3, Chap. 1.
15. Cf. *Capitalism, Socialism and Democracy*, op. cit., note 3, pp. 9–11.
16. Ibid., pp. 11–12.
17. Ibid., p. 12.
18. Ibid., pp. 12–13. In the closing paragraph of his chapter on 'The Civilization of Capitalism', Schumpeter commented: 'Things economic and social move by their own momentum and the ensuing situations compel individuals and groups to behave in certain ways whatever they may wish to do—not indeed by destroying their freedom of choice but by shaping the choosing mentalities and by narrowing the list of possibilities from which to choose. If this is the quintessence of Marxism then we all of us have got to be Marxists.' (Ibid., pp. 129–30.)
19. Cf. *Business Cycles*, Vol. 1, opl. cit., note 12, pp. 98–9.
20. The recent literature on rational expectations does not overcome Schumpeter's strictures here. The 'rationality' of rational expectations is limited by currently-available information, and thus the inherent uncertainty of the future is not eliminated.
21. 'Capitalism develops rationality and adds a new edge to it in two interconnected ways.

    First it exalts the monetary unit—not itself a creation of capitalism—into a unit of account. That is to say, capitalist practice turns the unit of money into a tool of rational cost–profit calculation, of which the towering monument is double-entry bookkeeping. Without going into this, we will notice that, primarily a product of the evolution of economic rationality, the cost–profit calculus in turn reacts upon that rationality; by crystallizing and defining numerically, it powerfully propels the logic of enterprise. And thus defined and quantified for the economic sector, this type of logic or attitude or method then starts upon its conqueror's career subjugating—rationalizing—man's tools and philosophies, his medical practice, his picture of the cosmos, his outlook on life, everything in fact including his concepts of beauty and justice and his spiritual ambitions. . . .

Second, rising capitalism produced not only the mental attitude of modern science, the attitude that consists in asking certain questions and in going about answering them in a certain way, but also the men and the means. By breaking up the feudal environment and disturbing the intellectual peace of manor and village (though there always was, of course, plenty to discuss and to fall out about in a convent), but especially by creating the social space for a new class that stood upon individual achievement in the economic field, it in turn attracted to that field the strong wills and the strong intellects.' (Cf. *Capitalism, Socialism and Democracy*, op. cit., note 3, pp. 123–4.)

22. It is also essential to Schumpeter's view, although the issues cannot be pursued here, that capitalist development *does* eventually subject the process of innovation itself to a high degree of rationality. The growth of the large corporation produces an environment within which innovation becomes the product of routinized activity—an activity involving a systematic search that utilizes scientific knowledge and methodology. However, this 'bureaucratization of the entrepreneurial function,' as Schumpeter refers to it, is associated with parallel social and political changes that are eventually responsible for the collapse of capitalist institutions. Cf. *Capitalism, Socialism and Democracy*, op. cit., note 3, Chaps. 12 and 13.
23. Cf. *The Theory of Economic Development*, op. cit., note 12, p. 65. (Emphasis Schumpeter's.)
24. Cf. *Business Cycles*, Vol. I, op. cit., note 12, p. 73.
25. 'Despite these qualifications, economic theory proceeds largely to take wants as fixed. This is primarily a case of division of labor. The economist has little to say about the formation of wants; this is the province of the psychologist. The economist's task is to trace the consequences of any given sets of wants. The legitimacy and justification for this abstraction must rest ultimately, as with any other abstraction, on the light that is shed and the power to predict that is yielded by the abstraction.' (Milton Friedman, *Price Theory*, New York, Aldine Publishing Company, 1976, p. 13.) Another reason economists have not considered it necessary to inquire into the determinants of changes in tastes is the assertion that the market behaviour of consumers can be adequately acounted for by reference to observable economic variables such as changes in prices and incomes. For a forceful formulation of this view, see George Stigler and Gary Becker, 'De Gustibus non est Disputandum', *American Economic Review*, **67** (1977), pp. 76–90.
26. Joseph A. Schumpeter, 'English Economists and the State-Managed Economy', *Journal of Political Economy*, **57** (1949), pp. 380–1, reprinted in *Essays*, op. cit., note 1, pp. 305–6.
27. It is also interesting to note that Schumpeter's views in these respects are further evidence of his commitment to an economic interpretation of historical change. Thus, if tastes do not stand as independent forces, but as social phenomena that are shaped by economic forces, they need to be regarded as part of the research agenda of economists. This is, of course, a position that would be emphatically rejected by neo-classical economists, but not by Marxists.
28. Cf. *Capitalism, Socialism, and Democracy*, op. cit., note 3, p. 77, fn. 5.

29. Ibid.
30. Joseph A. Schumpeter, 'Preface to Japanese Edition of "Theorie der wirtschaftlichen Entwicklung," ' reprinted in *Essays*, op. cit., note 1, pp. 158–63.
31. Cf. *Business Cycles*, Vol. 1, op. cit., note 12, Chap. 2.
32. 'Innovation is the outstanding fact in the economic history of capitalist society,' ibid., p. 86.
33. Cf. *History*, op. cit., note 3, p. 968.
34. Cf. *Capitalism, Socialism and Democracy*, op. cit., note 3, Part III.

10 Technological innovation and long waves: an inquiry into the nature and wealth of Christopher Freeman's thinking*

Luc Soete

Introduction

Christopher Freeman's contribution to the economics of technological change (as summarized in his seminal textbook *The Economics of Industrial Innovation*, 1974, 2nd edn, 1982) is generally identified with the microeconomics of industrial innovation, sometimes even more narrowly referred to as 'innovation studies'.[1] As many of the previous chapters have illustrated, Freeman's insights into the process of invention and innovation both at the firm and sector levels have indeed been invaluable. Apart from having pioneered research on the definition and international comparability of technological performance data,[2] Freeman's detailed perception of the process of technological innovation has probably been most valuable in clarifying many of the traditional economic misgivings, generalizations and simplifying assumptions[3] used in the microeconomics literature of technological change (including the areas of market structure and innovation, the determinants of the R & D system and the diffusion of technology).

However, the relevance of these innovation studies goes well beyond the rather artificial delimitation to microeconomics. This is probably best illustrated in the case of Freeman's work on innovativeness and research and development in the plastics[4] and electronic capital goods[5] industries, which has, along with the contributions of Posner, Hirsch and Hufbauer,[6] been instrumental in the 'rediscovery' of the crucial importance of the technology factor in international trade.

In this paper I shall briefly review some of Freeman's more recent, more macroeconomic and also, at first glance, more controversial work on the long-term employment implications of technological change and its relationship to so-called 'long waves'.[7] What will emerge from this review is the quite fundamental, path-breaking nature of Freeman's contribution despite the controversy surrounding this subject.

* In writing this review, I have greatly benefited from advice, comments, suggestions and editorial help from Jackie Fuller, Marie Jahoda, Roy MacLeod, Richard Nelson, Keith Pavitt and Nathan Rosenberg.

1. Innovation studies, macroeconomics and employment

The macroeconomics of technological change, it is probably fair to say, has traditionally been an area in which Freeman has felt least at ease. Apart from his deep-rooted apprehension about any macroeconomic aggregation it was precisely Freeman's detailed and intimate knowledge of the process of invention and innovation which made him so suspicious of the way technological change and particularly its diffusion had been introduced in macroeconomics. Research on the aggregate rate of technical change seemed to him if anything more to obscure the issue than to provide any useful insights.[8] Its purpose could in his view be traced back to the search for a simple representation of this cumbersome 'can of worms' variable, 'not to be opened except by scientists and engineers or occasionally by historians,'[9] which would fit in a more or less straightforward way the existing macroeconomic theories and tools of analysis, rather than question them in any way.

A typical example of this is the production function. Freeman's strong criticism of the production function as an analytical tool in the study of technological change and macroeconomic growth goes well beyond the traditional reservations about the unrealistic assumptions underlying the 'smooth' neo-classical version of the concept. For Freeman this concept:

> ignores the interrelatedness and complementarities of many technical and organizational innovations, thus tending to assume a spurious 'reversibility' and a wider spectrum of choice in technology, than exists in reality. It disregards the heterogeneity of many production inputs and the specificity of particular skills and types of production equipment. It ignores the extent to which particular 'technological trajectories' dictate the path of technical change for enterprises which aspire to remain competitive in international technological competition . . . it also ignores the firm-specific nature of much technical innovation and technological accumulation.[10]

Few 'technologically minded' economists would probably take issue with these views. Adherence to neo-classical production function analysis in this area is not one of well-founded 'faith'; if anything its use is often justified with reference to the lack of an alternative illuminating, if misleading, general equilibrium theoretical framework.[11] Present-day adherence to the concept of the production function in the area of the economics of technological change represents in many ways a clear recognition of the inherent limitations of most economic thinking in this area.

However, with respect to the employment implications of technological change—one of the oldest and most prolific issues of debate in economics—it is not so clear what the exact nature of the contribution of detailed innovation expertise and micro study could be. Worse, the question could be raised as to what extent micro-inspired innovation analyses might not represent in some way the 'mirror-image' of the macroeconomic aggregation problem raised above. Indeed, by pinpointing the direct micro occupational employment impact of a new technology (as much in terms of the resulting employment losses or employment growth), such micro studies, through their inherent methodological failure in fully comprehending the various macroeconomic employment compensation mechanisms, could be said to obscure rather than to clarify the issue. Many of these detailed technologically based studies in the 1950s and 1960s, with their early predictions of mass unemployment 'on a scale which would make that of the 1930s pale into insignificance,'[12] seem to have done just that.

There is little doubt that these early micro, technologically-inspired, often alarmist employment predictions have led many economists to approach micro-based innovation studies in this area with a great deal of scepticism. If anything, these studies, if not having killed the present debate, provide the strongest support ever for the crucial importance of macroeconomic employment compensation and the particular relevance of macroeconomic analysis in this area.

Some doubt as to the significance, if not relevance, of Freeman's contribution to this macroeconomic debate might therefore well be raised. Furthermore, to relate the technology-employment debate to an even more controversial topic such as 'long waves'—the subject of as many 'imaginary' as serious contributions—is prone to raise even more questions about the significance of Freeman's work.

Could it not be the case that Freeman, like so many other technology experts before him, has ignored many of the macroeconomic price adjustments leading to employment compensation, overestimating therefore the employment displacement nature of technological change? At the same time, is there also not a presumption that precisely because of his detailed technological knowledge, he will, as so many other innovation specialists, be inclined to overestimate both the revolutionary employment impact and the speed of the diffusion throughout the economy of these rather few *technically* revolutionary inventions or innovations?

I will argue here that this is *not the case*. On the contrary, Freeman's awareness of employment compensation, with its focus on the crucial role of *product* innovations, goes well beyond the traditional neo-classical partial equilibrium analysis of price and factor substitution employment compensation following the introduction of new process technologies. Similarly, in relation to the second question, it is precisely Freeman's intimate knowledge of the process of technological change and innovation, and in particular the conditions for commercial success and failure, which have led him to pay far more attention to the determinants of the adoption process and the major social and institutional, rather than just economic, bottlenecks in the widespread diffusion of radical innovations. It was indeed this factor which made him so careful about the timing of the growth and employment impact of 'clusters' of innovation and also brought him to reject in such an uncharacteristic way the simple 'innovation trigger' hypotheses,[13] popular in part of the long wave innovation literature.

Both these aspects of Freeman's understanding of the process of innovation and technological change are fundamental in correctly assessing his interpretation of technological change in affecting future employment. The rather common interpretation of Freeman's thinking in this area, as depicting a relatively pessimistic picture of significant future growth in unemployment as the result of technological change and its accompanying employment displacement,[14] is, in my view, based on a fundamental misunderstanding of Freeman's work.

Freeman's earlier, short-term pessimistic predictions[15] about the employment impact of technological change were, in the first instance, inspired by the ease with which the emerging structural employment crisis was being dismissed[16] as of the traditional recessive type (the result of an unfortunate external oil-shock crisis, to which labour markets would adjust in a relatively straightforward way) and, on a more personal level, by his deep-rooted human concern for the social and political consequences of mass unemployment. In other words, rather than ignoring employment compensation, the predictions of a significant rise in unemployment in much of Freeman's earlier analysis were based on a far more subtle and in the final instance more correct assessment of the importance of the time lags in labour market adjustment and macroeconomic employment compensation, than in many of the traditional macroeconomic analyses.

His more recent, long-term optimistic employment predictions do not contradict these earlier views. They are again inspired by his growing concern about the increased policy acceptance of large-scale unemployment and the fatalistic view which seems to characterize much recent research in this area.[17] For Freeman it is now crucial to indicate how technological change, through Schumpeter's 'perennial gales of creative destruction', will indeed lead to long-term output and employment growth; how the widespread diffusion and adoption of a 'cluster' of interrelated new technologies—Freeman's concept of 'new technological system'—will require organizational, social, institutional and economic reform; how indeed public policy has a fundamental role to play, if only in 'enabling' the emergence of this new 'socio-economic' paradigm.

Underlying Freeman's thinking in this area, one recognizes his fundamentally optimistic view of society's capacity for technological, economic, social and institutional adaptation which had already been forcefully set out in his critique of *The Limits to Growth*.[18] For Freeman, like Smith and Marx, technology is in the final instance a 'liberating force', one which has the potential for greater variety and satisfaction from work, one which could lead to 'higher degrees of autonomy, responsibility and skill within the workforce'.[19]

In the next sections I examine in somewhat more detail some of Freeman's ideas as they seem to have evolved over the last ten years. In the following section I will cover the period until 1978 when Freeman's ideas about the long-term relationship between technological change and employment were first moulded in a 'long wave' framework; in this period employment could be said to correspond to the main focus of analysis. In the third section I will discuss the period 1979–82, when the focus of analysis appears to have shifted to a more extensive and detailed investigation of the nature and timing of technological change. Finally, in the fourth section I will present the main features of Freeman's most recent writings—the period 1983–5. This most recent period is characterized by its broader, socio-institutional and organizational focus; the previous concept of the 'new technological system' is replaced by Perez's broader concept of the 'socio-economic' technological paradigm.

2. Employment and long waves: the early Freeman

I still remember vividly the day Freeman presented his 'Kondratiev' ideas for the first time in a seminar at the Science Policy Research Unit late in 1976. The impact he made on his listeners was truly phenomenal. Few people in the audience had ever heard of Kondratiev or the concept of long waves, although in continental Europe and particularly in Belgium and the Netherlands, both with strong traditions in the field of business cycles (Dupriez and Mandel in Belgium, Van Gelderen, De Wolff and Tinbergen in the Netherlands),[20] these ideas were far better known.[21]

After Freeman's seminar, many 'global modelling' researchers rushed to the computer to attempt to work out a precise formulation of a long wave model which would encompass the relationships he had described and which could be subjected to empirical testing. After a few months, with greater and greater difficulties encountered in quantifying the exact nature of the relationships and concepts put forward by Freeman, and the lack of any clear empirical evidence for the existence of long cycles, the interest of the modellers faded (with the notable exception of John Clark). The question was too difficult to be solved easily, too tricky to be formulated in a straightforward way, and too demanding in terms of the available data.

The episode illustrates well the complexity and the speculative nature of the 'long wave' debate within which Freeman formulated the relationship between technological change and employment. At the empirical level, there is indeed no statistically satisfactory evidence of a clear long wave behaviour in the economy.[22] At the theoretical level, on the other hand, there are at least a dozen competing explanations, ranging from purely monetary, to primarily demographic, political, institutional, technological or even philosophical arguments about the possible existence of long-term cycles.[23]

Freeman's contribution to this debate contains, however, a set of unique features. First and foremost, he introduced the issue of *employment* into the long wave debate. As Delbeke noticed: 'It is surprising that the long wave has only recently been studied by means of an explicit employment approach'.[24] While both Schumpeter, and to an even greater extent Lederer, did refer to employment in their analysis of the cyclical long-term development of capitalism and related this to the emergence and growth of 'innovative' firms,[25] the exact nature of the long-term relationship between technological change and employment had not been discussed. One reason for this was quite simply the unavailability of meaningful long-term data on employment and unemployment. For Freeman, typically, such data unavailability was of little relevance. In bringing employment into the long wave debate, Freeman broadened and made policy-relevant what had up to then been a purely academic debate among mainly European economic historians, about whether these 54.5-year cycles did indeed exist. As an outsider to this debate, Freeman was entitled to take a relatively eclectic stand, accepting both van der Zwan's[26] point that the concept of long cycles was really a relic of '19th Century socio-mechanics' and van Duijn's[27] various ingenuous attempts at providing.supportive evidence of the existence of innovation-related long waves. As he wrote in 1981: 'It is by no means necessary to accept the idea of cycles as such and certainly not the notion of fixed periodicity. Tinbergen prefers to speak of 'long waves' rather than 'long cycles', while van der Zwan prefers the notion of periodic major structural crises of adjustment and Mensch speaks of a 'metamorphosis' model. Almost all the advocates of the idea now prefer to speak of long waves rather than long cycles. For [our] purposes . . . it is necessary only to accept the fact that there have been some periods of "fast history" and some of "slow history" (to use Burton Klein's expression)'.[28]

This eclectic empirical approach allowed Freeman to develop his rather unique set of theoretical arguments concerning the differential employment effects of technological change over the various phases of the long wave. For Freeman it was clear that, in contrast to traditional macroeconomic growth theory, the technology–employment relationship would not remain constant over time. The long wave framework provided the perfect setting for a systematic and alternating technology employment-generating, technology employment-displacement impact. As Freeman put it:

> It seems inherently plausible that both the pressures and the opportunities for job-displacing technical change may vary over time . . . innovations may tend to be 'job-generating' in their early stages but 'job-displacing' as they mature. In the early stages there is no standardisation and often competing designs. Many new entrants come into the new industries and services attracted by the growth and profit prospects. Because of the absence of standardisation and of major economies of scale, many processes are highly labour-intensive, so that the job-generating effect may be considerable.
>
> However, as the new technology matures some standardisation takes place on the basis of the most successful designs and techniques; production and distribution economies of scale become increasingly important; firm size and capital intensity tend to increase. Employment grows more slowly or stabilizes. In some cases secondary 'job-displacement' effects in other industries and services may become very important. . . .
>
> Such a perspective does demonstrate the long time-lags involved in the introduction and spread of major new technologies and does therefore lend greater credence to Kondratiev's concept of long waves in economic development, although not to pedantic notions of their precise timing.[29]

In emphasizing the 'job-displacing' nature of technological change in periods of 'slow history', Freeman, like so many other recent long wave converters, was obviously much influenced by the severity of the 1975 recession and the rapid increase in unemployment over 1975–6. In contrast to most macroeconomists, it was obvious to Freeman that the traditional macroeconomic demand remedies were totally inadequate in coping with the growing unemployment problem. It is probably fair to say that Freeman felt in many ways compelled to write a macroeconomic employment paper,[30] because of the need to point to the fundamental structural nature of the emerging employment crisis and to warn about the long-term employment problem most OECD economies would be facing. However, long before 1975–6, the time of writing of his first major paper on the subject, he had become concerned about the gradual increase in 'structural' unemployment with each new business cycle.

Paradoxically, in drawing his pessimistic employment–technology conclusions, Freeman was undoubtedly far more careful in his assessment of the adjustment and assimilation lags of major radical technologies than many macroeconomists were in relation to price and labour market adjustments. For Freeman:

> The technologists and the computer men may have been right in the 1950s in recognizing the enormous labour-displacing potential of the electronic revolution, but wrong in failing to recognize the social and economic time-lags involved in such an upheaval and the further improvements in technology which were needed. Even in the 1960s, the installation of computers was usually associated with the generation of many new jobs, rather than with job displacement, and there were still acute shortages of systems analysts, programers and other computer personnel. . . . Only by the 1970s had all these processes reached a point when job displacement effects began to be more acutely felt.[31]

Freeman's belief in the existence of long-term phases in economic development, while never explicitly stated before 1976–7, goes back, however, to the pre- and post-war period when, as a student at the London School of Economics (LSE), he was first confronted with the growth retardation debate. The question of whether the post-war period would see a continuation of the severe 1930s Depression was an issue in many quarters. The economic boom of the late 1930s and 1940s seemed very much 'war related'; once the reconstruction 'boom' of the early post-war period petered out, there seemed little reason to expect continued economic expansion. It is also worth remembering that at around the same time (1947–53) there was a similar debate in the Soviet Union (at the Institute of World Economy) about the long-term decline of capitalism, some of which had made some impact in the West (in particular Varga's contributions).[32] Following Varga, Freeman believed in the existence of an eight to eleven-year 'reproduction cycle' which would last until about 1958. When, in 1959, economic growth appeared to resume practically uninterruptedly, the assumption of a longer-term economic development pattern on which these Kuznets or Marxian cycles could be 'grafted' began to appeal more and more to him.

However, it is probably fair to say that these early 'thoughts' and impressions were not based on any systematic understanding of the nature, the determinants and the relationship between such long-term cycles and the emergence of radical new technologies. To find such a first attempt one had to wait until the mid-1970s, and the emergence of what I have typified here as the 'early Freeman'.

3. New technological systems: the Schumpeterian Freeman

In Freeman's early writings, the role of technological change had been primarily confined to its assumed effects on both the displacement and the generation of new jobs. While Freeman acknowledged the implicit measurement and conceptual problems in identifying these effects, he related the latter in a more or less straightforward way to the quite distinctive effects of process as opposed to product innovations. As he put it:

> It is clear that the process of economic growth does involve two distinct processes: one of *job displacement* and the other of *job generation*. The rate of *job displacement* will depend to a considerable extent on the technical effort which is devoted both within and outside professional R & D laboratories to labour-saving innovations and of course on the purely scientific and technical possibility of achieving such economies. It will also depend on the attitudes of management and labour, and the perceived advantages and disadvantages of adopting labour-saving techniques. The rate of *job generation* on the other hand will depend to a considerable extent on perceived new opportunities for profitable investment. These will depend at least in part on the full exploitation of radical innovations in new fields, leading to the creation of new industries and services, such as (in their day) automobiles, railways, radio, television or drugs.[33]

The period 1979–82 however, signalled, a gradual shift in Freeman's writings away from this long-term 'employment' focus and its link with product and process innovations, towards the broader issue of the relationship between innovation and long-term economic fluctuations. This led him to specify in a more detailed way the nature of these *radical* innovation 'bunches' or clusters, and in particular their timing, their determinants and process (as opposed to product) impact. The widening of his analysis also brought him far closer to Schumpeter's work. Overall, while this period undoubtedly marked a shift away from the long-term employment technology relationship, at a time when the issue was getting more policy recognition both in academic and policy-orientated circles, the period 1979–82 saw a 'Schumpeterian Freeman' refine and expand his theory in a number of quite distinctive directions.

First and foremost, Freeman put forward the concept of 'new technological systems', going well beyond the simple concept of 'clusters' of radical innovations used before. As he put it:

> We are interested in 'constellations' of innovations which have a relationship to each other and not just in the more or less accidental statistical grouping of the innovation of a particular year or decade. The important phenomenon to elucidate if we are to make progress in understanding the linkages between innovations and long waves is the birth, growth, maturity and decline of *industries and technologies* (pp. 64–65). . . . Thus we are interested in what we shall call 'new technology systems' rather than haphazard bunches of discrete 'basic innovations'. From this standpoint, which we believe was essentially

that of Schumpeter, the 'clusters' of innovations are associated with a technological web, with the growth of new industries and services involving distinct new groupings of firms with their own 'subculture' and distinct technology, and with new patterns of consumer behaviour.[34]

While there was no overall attempt at defining, let alone measuring, the concept of 'new technological system', Freeman nevertheless succeeded in illustrating the latter in a particularly neat way in the case of the emergence and growth of the plastics and electronics industries.

In developing his concept of 'new technological systems' and explaining the 'clustering' hypothesis, Freeman (in contrast to many other long wave writers) was at pains to emphasize the crucial importance of breakthroughs and new developments in fundamental science, rather than the simple clustering of basic innovations:

Thus, for example, the growth of the synthetic materials industry was not just a question of a bunch of discrete product innovations, although this did of course occur; it was also a question of the development of a common technology (polymer science) and of a whole series of process innovations and improvements, and of applications innovations to meet the needs of scientific applications, of plasticiser and emulsifier innovations, of innovations in machinery, such as injection moulding machines.[35]

As Chesnais[36] observed in one of the most thorough reviews on the subject, alongside the particular emphasis on basic science as a 'trigger' for a possible 'clustering' of basic innovations, one of the most distinctive features of Freeman's thinking in this area, going well beyond Schumpeter, was the importance given to wars and war preparations as a crucial 'demand-pull' innovation clustering factor. Indeed, and again building to a considerable extent on his detailed 'innovation' knowledge, Freeman went a long way to stress the interaction between these basic science 'supply-push' and wartime 'demand-pull' factors:

We are certainly not advocating a pure 'science-push' theory of innovation. In our view, most case study work on the history of innovation, and especially Project Sappho [led by Freeman], point to the interdependence of 'supply' and 'demand' factors and to the importance of a favourable combination of circumstances. Therefore we would not underestimate the significance of demand, whether actual or anticipated, or that of entrepreneurial attitudes and behaviour. The German chemical trust IG Farben accounted for an extraordinary high proportion of the world's synthetic materials innovations and, in our view, its success was related both to the strength of its R & D, including its strong links with basic research through its consultancy arrangements with

> Staudinger and other outstanding chemists, and to the peculiar nature of German demand, strongly influenced by aspirations to overcome dependence on imported natural materials.[37]

From this it will be clear that the concept of a 'new technological system' put forward by Freeman represents a more complex and undoubtedly richer framework of thinking than Schumpeter's relatively simplistic view of innovation clustering—the result of the non-randomness of innovations, and of their 'lop-sided, discontinuous and disharmonious' nature—and his rather uni-directional theory about the effect of these major technological discontinuities on investment and employment. It remains nevertheless fair to say that the inspiration for Freeman's more complex 'clustering' hypothesis is by and large Schumpeterian in nature. Indeed, for Freeman his theory 'does not mean abandoning the Schumpeterian notion of a reciprocal association between innovation and long-term economic fluctuations. But the nature of this relationship is a good deal more complex and untidy than a simple clustering of a large number of 'basic innovations' in particular decades every half century or so'.[38]

Second, and closely related to the idea of 'new technological systems', technology *diffusion*, including Rosenberg's *proviso* about the importance of incremental innovations over the diffusion phase, now became a crucial part of Freeman's long wave innovation theory. Whereas in his previous work Freeman had emphasized the time-lags involved in any factor-saving technology inducement mechanism,[39] he had barely pursued fully the importance of the technology diffusion factor in the discussion of these time-lags, at one time even appearing to accept Mensch's argument in relation to a 'dearth of radical innovations' as 'possible explanation of flagging investment'.[40] In his later work, however, Freeman fully acknowledged the importance of the technology diffusion factor and, emphasizing further his Schumpeterian allegiances, made the 'competing away' of innovation based 'quasi-rents' over the diffusion phase the centrepiece of his analysis.

In shaping this more formally Schumpeterian, 'time-lag' aspect of Freeman's thinking, the influence of Stan Metcalfe's contribution to the diffusion debate is essential. Metcalfe,[41] in a seminal contribution, attempted to integrate a number of supply factors in the primarily demand-dominated diffusion models. According to Metcalfe:

> The pace of diffusion depends on supply constraints just as much as it does on the traditional demand constraints of adoption. Even if uncertainty considerations did not delay adoption, diffusion would still not be instantaneous—as the standard model would predict—but would depend upon the pace at which productive capacity could be built up.[42]

In doing so, Metcalfe emphasized the link between such a more complete diffusion picture and the early structural change literature.[43]

> Diffusion of innovation is one dimension of that vitally important but not well understood problem of the transition between different economic equilibria. It represents the study in the small of dis-equilibrium situations, of Schumpeter's process of creative destruction. From this perspective, compelling links can be made between the micro-study of diffusion phenomena and the general process of modern industrial growth, in which the impulses associated with technological innovations are absorbed into the economic structure.[44]

Whereas Metcalfe did not go as far as suggesting a link between his more comprehensive supply–demand diffusion model, the growth retardation debate and the possible existence of long waves, Freeman would be quick to acknowledge such a possible link.

Third, his emphasis on 'new technological systems' and the 'diffusion', rather than the mere occurrence of 'basic' innovations, led him to put forward a quite severe critique of the simple, depression-induced, innovation-bunching and invention-innovation accelerating hypothesis popular in much of the technology-inspired long wave literature, and associated in particular with the name of Mensch.[45] For Freeman there is neither theoretical nor empirical support for Mensch's hypotheses.

First, and again very much in line with Schumpeter, it is not so much the individual dating of particular innovations which matters to Freeman as much as the diffusion and subsequent swarming effects of those innovations:

> What matters in terms of major economic effects is not the date of the basic innovation; what matters is the diffusion of this innovation—what Schumpeter vividly described as the 'swarming' process when imitators begin to realise the profitable potential of the new product or process and start to invest heavily. This swarming may not necessarily occur immediately after a basic innovation although it may do so if other conditions are favourable. In fact, it may often be delayed for a decade or more until profitability is clearly demonstrated or other facilitating basic and organisational innovations are made, or related social changes occur. Once swarming does start it has powerful multiplier effects in generating additional demands on the economy for capital goods, for materials, components, distribution facilities, and of course labour. This, in turn, induces a further wave of process and applications innovations. It is this combination of related and induced innovations which gives rise to expansionary effects in the economy as a whole.[46]

Second, Mensch's hypothesis about the bunching of unrelated basic innovations again amounts in Freeman's view to a far too simplistic, near mechanical view of the relationship between technological innovation and economic recovery. For Freeman:

> Mensch has been looking at the wrong 'swarms'. . . . Surprisingly when [Mensch] speaks of the 'bandwagon' effect in relation to a cluster of basic innovations, he is apparently not talking about Schumpeter's 'one-sided rushes' of firms anxious to jump in and get above-average profits in a rapidly growing new branch of industry, he is talking about a disparate set of basic innovations. It is very hard to see in what sense the originally quite separate launch of helicopters, television, tetraethyl lead, titanium, etc. in the mid-1930s could constitute a 'bandwagon' in any normal meaning of the term. The swarms which matter in terms of their expansionary effects are the diffusion swarms *after* the basic innovations and the swarming effects associated with a set of inter-related basic innovations, some social and some technical, and concentrated very unevenly in specific sectors.[47]

Third, and probably most clearly acknowledged in the more recent long wave literature,[48] Freeman illustrates case by case how the evidence gathered by Mensch in support of his so-called 'acceleration' hypothesis is unconvincing and rests upon a rather artificial 'selection' of a number of innovations from the sixty or so case studies discussed by Jewkes *et al.*[49] Again, I would argue that it was ultimately Freeman's detailed knowledge of the process of innovation and its major determinants which allowed him to rebut so clearly the core of Mensch's 'metamorphosis' model.

4. Technological paradigms and institutional change: the rejuvenated Freeman

The emergence of Freeman's previous concept of 'new technological systems' might well be considered as similar to the concept of change in technological paradigm, already used extensively in his Schumpeterian writings.[50] However, until 1983 the 'paradigm' concept was only introduced in Freeman's long wave thinking in the more narrow 'technological' meaning of the term, as developed by Dosi in his now classic contribution.[51] Furthermore, whereas the change in 'paradigm role' of a 'new technology system' was indeed already acknowledged (in these relatively narrow technological terms) in the 'Schumpeterian Freeman', the emphasis was on the application and exploitation over the various phases of the long wave of the broad, Nelson and Winter type, 'natural trajectories' of technology, such as mechanization, electrification or automation.[52] To put it in Dosi's terminology, the focus appeared more on the 'thermostat' function of technical change, rather than on its 'engine' function. This was in many ways understandable: much in line with his earlier thoughts, Freeman's concerns, even though they had shifted, were still very much geared towards an explanation of why over the various phases of the long wave technical change would change in its employment growth effects and with the onslaught of the recession might become predominantly labour-saving.

The emergence of what I have characterized here as the 'Rejuvenated Freeman' amounts therefore to a new episode in Freeman's thinking on the subject of technical change, employment and long waves. The most distinctive features of this recent period can be summarized as follows: first and foremost, Freeman has broadened his still relatively narrow, technology-orientated concept of 'new technological systems' to the far broader concept, suggested by Perez[53] in one of the most influential articles in the area of long waves,[54] of change in 'technological regime' or change in 'techno-economic' or 'socio-economic paradigm'.

The view that the structural crisis brought about by the depression is, in Perez's words, 'the visible syndrome of a breakdown in the complementarity between the dynamics of the economic subsystem and the related dynamics of the socio-institutional framework' and amounts not only to 'a process of "creative destruction" or "abnormal liquidation" in the economic sphere, but also in the social-institutional,'[55] appeals particularly to Freeman. Indeed, Freeman, despite his strong Schumpeterian allegiances, fully acknowledges the point made by Perez that Schumpeter did only have a narrow economic (if any) interpretation of the occurrence of depressions. As he puts it: 'Despite [Schumpeter's] acceptance of the importance of organisational and managerial innovations and the breadth of his approach to the development of social systems, his theory of depression is narrowly economic', and further reiterating the point made by Perez, he adds: 'But it is the "mismatch" between the institutional framework with its high degree of inertia, and the outstanding revealed cost and productivity advantages of the new technological paradigm which provides the impulse to search for social and political solutions of the crisis.'[56]

The second most distinctive feature with which I would identify this recent period—a change of emphasis towards the long-term growth and employment-generating implications of the new emerging technological paradigm—is probably more controversial and has so far only been acknowledged in some quarters. However, as mentioned in the second section, it is really in the logic of Freeman's long wave thinking that, with the further deepening of the depression and the historical record levels in unemployment in all OECD countries, the focus of his analysis would gradually shift. From an emphasis in his earlier writings on the predominantly labour-saving biases linked with rationalization investment and automation and the time-lags involved in the resulting employment compensation effects, he now points increasingly to the long-term growth and employment potential of the new, emerging technological paradigm.

With employment forecasts becoming more and more pessimistic, and with questions being raised about the long-term potential of employment creation of Western societies,[57] it becomes essential to Freeman to change the emphasis in his writings and to illustrate how recovery will bring about employment growth: not by simply reiterating the general principles of neo-classical compensation theory or denying the new technology its revolutionary character, but by drawing the logical growth and employment implications of the new 'techno-economic paradigm' identified with Information Technology. Again, Perez's[58] influence is fundamental. She indeed stresses to Freeman the far more widespread and radical shift in all aspects of organizational and social life implied by the emergence of the new 'cheap electronics' paradigm. As written in her joint paper with Freeman:

> Today, with cheap microelectronics widely available, and consequently low cost information handling, energy and materials intensity is no longer 'common sense'. The ideal productive organisation brings together management and production into one single integrated system for turning out a flexible output of preferably information-intensive, rapidly changing, products and services. . . . The skill profile changes from middle range to increasingly high and low range qualifications, and from narrow specialisation to broader and multipurpose basic skills for information handling. Diversity and flexibility at all levels substitute identity and massification as 'common sense' best practice.[59]

Combined with his detailed knowledge of the process of technological innovation, Freeman is now able to put forward one of the most thorough accounts of the characteristics and distinctive features of the 'new information technology paradigm',[60] bringing out at the same time the fundamental difference with the concept of 'automation', and drawing a fascinating analogy with the emergence of electric power in the late nineteenth and early twentieth century.

Both previous features will bring Freeman to hammer out, in a far more integrated and comprehensive way, the policy implications of his thinking. Indeed, Freeman now brings out the importance of the required need for change transcending the purely technological or economic field. As Perez puts it, in her joint contribution with Freeman: 'depression is a "shouting" need for full-scale reaccommodation of social behaviour and institutions in order to suit the requirements of a major shift which has already taken place in the techno-economic sphere.'[61] The extent of the required social, organizational and institutional change is vast and includes practically all facets of economic life: the education and training system, the industrial relations system, the prevailing production and management 'styles', the financial system, and the international trading and monetary systems.

The policy implications are self-evident. They include many aspects which were already present in Freeman's previous writings—such as the need for a generally more educated work-force and for a more explicitly formulated innovation policy. A crucial point is, however, that these more micro-orientated policy prescriptions now become part of a far more cohesive and comprehensive set of 'meso' policy recommendations, with the central emphasis on the need for institutional change. Indeed, to a far greater extent than, for instance, Schumpeter, institutional change now becomes the cornerstone of any new recovery.

While Freeman remains undoubtedly less outspoken about the employment implications of such a recovery, he is nevertheless at pains to stress the various complementarities of the 'virtuous circle' of rapid growth: high productivity gains, high profitability and the creation of many new employment opportunities. But the extent of social, organizational and institutional change required to move to such a new growth path and the challenge it presents to society's capacity for change is formidable indeed. As he puts it in his rather unique way:

> The institutional framework for future economic recovery is only now being shaped. Most of our institutions and ideologies are still geared to the old post-war technological paradigm. Only through social and political debate and conflict shall we determine how we reshape our institutions and our way of life to match the potential of the new technology and to humanize its innumerable potential applications. The new patterns of employment which will emerge should be of a kind which encourage great variety in hours of work and in continuing education and training, but which ensure to everyone who is seeking paid employment the opportunities to work in socially useful activity. This should mean a renewed commitment by society to the goal of 'full employment', but in a new social context which takes acount both of changes in technology and changes in society.[62]

Conclusions

In this brief view, I have tried to illustrate the importance and 'evolutionary' logic of Freeman's contribution to the debate about the employment implications of technical change over the last ten years. That contribution, as much as its 'evolutionary trajectory', is far from over. In ending this review let me therefore list some of the directions in which

I would particularly like to see Freeman making further substantial inroads in our ignorance of the long-term relationship between technological innovation, economic growth and employment.

First, and following directly from the previous section, there is probably a need to describe more explicitly the policy implications implicit in the concept of 'institutional change'. As it stands, the concept carries with it a certain vagueness. There is little doubt that to come up with a full list of the 'required' institutional changes would directly undermine the fully endogenous interaction between the techno-economic sphere and the socio-institutional framework implicit in Perez's concept of new technological paradigm. The process of social experimentation upon which we are, in Freeman's view, now embarked is indeed fully part of the further spread of the new 'technological regime'. But it should at least be possible on the basis of what we know already and what Freeman has spelled out so masterfully with respect to the new information technology paradigm to come up with some guidelines as to the required institutional changes in most areas.

Indeed, if the social-institutional framework is to be more than just a passive environment, but a causal factor in the explanation of long waves, one would imagine that a detailed description of the required institutional change would become essential for the new-found policy relevance of long wave theory. Some policy implications with respect to work organization have actually been set out in some detail by Freeman.[63] But in many other areas, where one would assume that the need for institutional change might be as dramatic (the present trade union structure, the telecommunications sector monopoly in most OECD countries, privatization and the role of public corporations, regional policy, etc.), Freeman has not yet given any indication of what the policy implications of the new 'information paradigm' might be.

Second, Freeman's recent emphasis on the socio-institutional framework as a crucial factor in creating the 'matching' conditions with the new techno-economic paradigm, should offer plenty of scope for closer integration with the so-called 'institutional' long wave theories, identified particularly with Gordon, Edwards and Reich,[64] and with the French 'regulation' school identified with the names of Aglietta, Lipietz, Boyer and Mistral.[65] So far, Freeman has made little reference to these authors. It is true of course that, with the notable exception of Dosi,[66] none of these authors have looked in such a systematic way or brought out so clearly the mutual interaction between the newly emerging techno-economic paradigm and the social-institutional framework encompassed in Perez's concepts of 'good match' and 'mismatch' used by Freeman. Some of these contributions could, however, particularly in their historic content, shed some further light on the extent to which these institutional 'matches', characteristic of the upturn and boom period of the long wave, emerged only after growth was already set in motion as a result of the emergence and further diffusion of the set of new radical technologies, or the other way round. Indeed, is the socio-institutional framework an *enabling* factor or a *determining* factor for future economic recovery?

Third, despite the empirical difficulties of finding supporting evidence of long wave behaviour, let alone the existence of causal mechanisms, there should nevertheless be scope for a more systematic empirical analysis of the emergence and further diffusion of Freeman's 'new technological systems' or new 'techno-economic paradigms', particularly in their economic (both macro and sectoral) impact and organizational and institutional 'match'. That evidence could bring out more clearly than before some of the remaining fundamental questions about long wave *theory* as put forward in Rosenberg and Frischtak's excellent but critical assessment of the debate. I would in particular refer to the 'recurrence' argument implicit in any long wave theory. Neither in the 'Early' nor in the 'Schumpeterian' Freeman was there any attempt to argue in terms of every half century seeing recurring clusters of new innovations or 'new technological systems'. Freeman's long wave theory was typically 'eclectic'—the notion of 'long wave' providing more an interesting dynamic disequilibrium concept, serving more as a useful historical framework than as a full-blown theory. With the concept of the 'new techno-economic paradigm', some notion of 'recurrence' is introduced, but its timing is as yet unclear, and in need of empirical evidence.

Fourth, the question could be raised as to whether there is really a need to mould Freeman's theory, and particularly the most recent concept of new 'techno-economic paradigm', into a long wave framework. It is clear that Freeman *believes* in long waves, but if both theory and supporting evidence remain on too shaky ground, is there not a danger of failing to convince any except the already converted? Is there not a danger of one of the most important theoretical contributions to the intricate relationship between technological innovation, structural change and long-term economic growth being discarded or ignored by many economists and policymakers?

Time will tell. Freeman has so far provided us with one of the most fascinating accounts of the interaction between the process of technological innovation, employment and the dynamics of capitalist economies. There is, however, much more to come. The reader of traditional Festschrifts be warned: rather than having acquired the final 'honouring' reference book on Freeman, start watching out for Christopher Freeman.

Notes

1. See, e.g., Z. Griliches (ed.), *R & D, Patents and Productivity*, Chicago/London, University of Chicago Press, 1984, referring to C. Freeman, *The Economics of Industrial Innovation*, Harmondsworth, Penguin Modern Economic Texts, 1974, 2nd edn, London, Frances Pinter, 1982; also Spanish edn, *La Teoria Economica de la Innovación Industrial*, Madrid, Penguin Alianza, 1975.
2. See, among others, Freeman's various reports on the measurement of Research and Development for the OECD: C. Freeman, *Proposed Standard Practice for Surveys of Research and Development: The Measurement of Scientific and Technical Activities*, Paris, OECD, Directorate of Scientific Affairs, 1962; C. Freeman, R. Poignant and I. Svennilson, *Science, Economic Growth and Government Policy: Background Paper for Ministerial Meeting on Science, Agenda Item III*, Paris, OECD, 1963; C. Freeman, A. Young, *The Research and Development Effort in Western Europe, North America and the Soviet Union: An Experimental International Comparison of Expenditure and Manpower in 1962*, Paris, OECD, 1966, culminating in his writing the Frascati Manual, OECD, *The Measurement of Scientific and Technical Activities*, Frascati Manual, DSTI, Paris, 1963 and subsequent revisions 1970, 1976 and 1981—which could more appropriately be referred to as the Freeman Manual; his reports for UNESCO on both the measurement and output of R & D: C. Freeman, *The Measurement of Output of Research and Experimental Development: A Review Paper*, Paris, UNESCO, 1969; C. Freeman, *The Measurement of Output of Research and Experimental Development: A Review Paper*, Paris, UNESCO, 1969; C. Freeman, *The Measurement of Scientific and Technological Activities*, Paris, UNESCO, 1969; his earlier reports for the Federation of British Industry on industrial research in manufacturing: C. Freeman, C. T. Saunders and R. W. Eveley, *Industrial Research in Manufacturing Industry, 1959–60*, London, FBI, 1961, and his report to the Bolton Commission: C. Freeman *et al.*, *The Role of Small Firms in Innovation in the United Kingdom Since 1945*, Committee of Inquiry on Small Firms, Research Report No. 6, HMSO, 1971.
3. As Rosenberg points out: 'With apologies to Clemenceau it might be said that if technological change is not too important a subject to be left to the economist, it certainly is too diverse a subject to be left to the economist who refuses to step across narrow disciplinary boundaries.' N. Rosenberg, *Perspectives on Technology*, Cambridge, Cambridge University Press, 2nd edn, 1985, p. 1.
4. C. Freeman, 'The Plastics Industry: A Comparative Study of Research and Innovation', *National Institute Economic Review*, No. 26, (1963), pp. 22–49.
5. C. Freeman, 'Research and Development in Electronic Capital Goods', *National Institute Economic Review*, No. 34 (1965), pp. 40–91.
6. M. Posner, 'International Trade and Technical Change', *Oxford Economic Papers*, **13** (1961), pp. 323–41; S. Hirsch, 'The US Electronics Industry in International Trade', *National Institute Economic Review*, No. 34 (1965), pp. 92–7; G. Hufbauer, *Synthetic Materials in the Theory of International Trade*, London, Duckworth, 1966.
7. The extent of the controversy is probably best illustrated by the scepticism with which his ideas have been greeted by some of his closest colleagues. For example, in discussing the question of the relationship between rapid innovation and employment, Nelson reads Freeman 'as schizophrenic': R. Nelson,

'Comment on Chesnais' in H. Giersch (ed.), *Emerging Technologies: Consequences for Economic Growth, Structural Change and Employment*, Tübingen, J. C. B. Mohr 1982. See also Pavitt's contribution to this Festschrift or even Rosenberg's excellent, but highly critical review of the technology-long wave issue (N. Rosenberg and C. R. Frischtak, 'Long Waves and Economic Growth: A Critical Appraisal', *American Economic Review*, Papers and Proceedings, **73**, No. 2 (1983), pp. 354–61 and N. Rosenberg and C. R. Frischtak, 'Technological Innovation and Long Waves', *Cambridge Journal of Economics*, **8**., No. 1 (1984), pp. 7–24.

8. As Freeman *et al.* put it: 'the way most of these questions are put (i.e in terms of general equilibrium) makes them of very little use in explaining major fluctuations as it is implicitly assumed that departures from the equilibrium path are short-lived frictional imperfections. A steady growth path, a dynamic full employment general equilibrium, a constant rate of 'Harrod-neutral' technical change are all concepts, which—at a certain stage—seem more to obscure than to clarify the actual interactions between growth, technical change and employment.' (C. Freeman, J. A. Clark and L. L. G. Soete, *Unemployment and Technological Innovation: A Study of Long Waves and Economic Development*, London, Frances Pinter, 1982, p. 31.)
9. C. Freeman, 'Innovation as an Engine of Economic Growth: Retrospect and Prospects' in H. Giersch (ed.), *Emerging Technologies: Consequences for Economic Growth, Structural Change and Employment*, proceedings of Kiel Symposium 1981, Tübingen, J. C. B. Mohr, 1982, pp. 1–32.
10. C. Freeman and L. Soete (eds), *Technical Change and Full Employment*, Oxford, Basil Blackwell, 1986, Chap. 3.
11. This is not to deny the importance of some of the alternative formulations of production relationships and their change over time. It is interesting to notice that most of these, starting from the earliest attempts at 'embodying' the transmission mechanism of technological progress rather than the rate and bias of technological progress (i.e. still assuming a totally exogenous rate of technological progress, such as Solow's vintage production formulation: R. Solow, 'Investment and Technical Progress', in K. Arrow, S. Karlon and P. Suppes (eds), *Mathematical Methods in the Social Sciences*, Stanford, Stanford University Press, 1960, and Salter's 'putty–clay' vintage model: W. Salter, *Productivity and Technical Change*, Cambridge, Cambridge University Press, 1960 to the more 'endogenous' formulations attempts of technological progress (such as Kaldor's 'technical production function' concept: N. Kaldor, 'A Model of Economic Growth', *Economic Journal*, **71** (1962), pp. 591–624; N. Kaldor, 'Capital Accumulation and Economic Growth' in F. Lutz and D. Hague, *The Theory of Capital*, London, IEA, 1965; Kennedy-Weizsäcker's 'invention possibly frontier' and induced invention hypothesis: C. Kennedy, 'Induced Bias in Innovation and the Theory of Distribution', *Economic Journal*, **74** (1964), pp. 541–7; C. von Weizsäcker, 'Tentative Notes on a Two-Sector Model with Induced Technical Progress', *Review of Economic Studies*, (1966), pp. 245–51; Arrow's 'learning-by-doing' model: K. Arrow, 'Economic Welfare and the Allocation of Resources for Invention', in National Bureau of Economic Research, *The Rate and Direction of Inventive Activity*, Princeton, Princeton University Press, 1962; Atkinson and Stiglitz's alternative formulation of technological change: A. Atkinson and J. Stiglitz, 'A New View of Technological Change', *Economic Journal*, **78** (1969), pp. 573–8; or David's concept of 'localized technological change': P. A. David, *Technical*

Choice, Innovation and Economic Growth, Cambridge, Cambridge University Press, 1975) did indeed emerge from a fundamental dissatisfaction with the way *technology* had been translated in the concept of the production function and neo-classical growth models. It is fair to say, however, that most of these, with a few notable exceptions such as the contribution of Atkinson and Stiglitz referred to above, were more concerned with the search for the exact conditions under which 'Golden Age' balanced growth equilibrium would still prevail in the economy rather than, as Cooper and Clark put it, with 'the question of whether technological change has systematic effects on the stability of economic growth': C. M. Cooper and J. Clark, *Employment, Economics and Technology: The Impact of Technical Change on the Labour Market*, Brighton, Wheatsheaf, 1982, p. 62.

12. See Wiener's *Human Use of Human Beings*, New York, Houghton Mifflin, 1949.
13. See in particular the contribution of G. Mensch, *Das Technologische Patt; Innovation überwinden die Depression*, Frankfurt, Umschau, 1975 and *Stalemate in Technology: Innovations Overcome the Depression*, New York, Ballinger, 1979.
14. See, e.g., Missika's review of the debate: J. L. Missika, O. Pastré, O. Meyer, J.-L. Truel, R. Zarada and C. Stoffaes, *Informatisation et Emploi: Menace ou Mutation?, informatisation et societé* **11**, Paris, La Documentation Française, 1981).
15. I refer here in particular to Freeman's contributions of the late 1970s: e.g., C. Freeman, 'The Kondratiev Long-Waves, Technical Change and Unemployment', paper prepared for OECD Experts' Meeting on Structural Determinants of Employment and Unemployment, Paris, March, 1977, published in *Structural Determinants of Employment and Unemployment*, Vol. 2, reports prepared for Experts' Meeting in Paris, 7–11 March 1977, Paris, OECD, 1979, pp. 181–96; C. Freeman, 'Technical Change and Unemployment', paper presented to Conference on 'Science, Technology and Public Policy: An International Perspective', University of New South Wales, 1–2 December 1977; C. Freeman, 'Technology and Employment: Long Waves in Technical Change and Economic Development', Holst Memorial Lecture, Eindhoven, The Netherlands, December 1978; C. Freeman, 'Technical Change and Unemployment' in S. Encel and J. Ronayne (eds), *Science, Technology and Public Policy: An International Perspective*, Rushcutters Bay, Pergamon Press, 1979, pp. 53–76; C. Freeman, 'Microelectronics and Unemployment' in *Automation and Unemployment*, papers presented at an ANZAAS Symposium, Sydney, The Law Book Co. Ltd., 1979, pp. 99–113; and C. Freeman, 'Unemployment and Government' in T. Forester (ed.), *The Microelectronics Revolution: the Complete Guide to the New Technology and its Impact on Society*, Oxford, Basil Blackwell, 1980, pp. 308–17.
16. In particular the conclusions of the so-called McCracken report, *Towards Full Employment and Price Stability*, Paris, OECD, 1977.
17. The concept of a 'natural' rate of unemployment which would have risen rather dramatically over the last decade was particularly offensive to Freeman.
18. See in particular his contributions to C. Freeman, H. S. D. Cole, M. Jahoda, K. L. R. Pavitt (eds), *Thinking about the Future: A Critique of the 'The Limits to Growth'*, London and Brighton, Chatto and Windus/Sussex University Press, 1973; also published in *Futures*, **5** March and April 1973, American edition: *Models of Doom: a Critique of 'The Limits to Growth'*, New York:

Universe Books, 1973; German edition: *Zukunft aus dem Computer? Eine Antwort auf 'Die Grenzen des Wachstums'*, Munich, Luchterhand, 1973; French edition: *L'Anti-Malthus: une Critique de 'Halte à la Croissance'*, Paris, Seuil, 1974.

19. Freeman, op. cit., note 10.
20. See also L. H. Dupriez, *Des Mouvements Économiques Généraux*, Louvain, 1947; E. Mandel, *Long Waves of Capitalist Development: The Marxist Interpretation*, Cambridge, Cambridge University Press, 1980; J. van Gelderen, 'Springvloed: beschouwingen over industriële ontwikkeling en prijsbeweging', *De Nieuwe Tijd*, **18**, (1913); S. de Wolff, *Het Economisch Getij*, Amsterdam, Emmering, 1929; J. Tinbergen and J. J. Polak, *The Dynamics of Business Cycles*, Chicago, Routledge and Kegan Paul, 1950.
21. Long waves or long cycles ideas refer to the apparent tendency for long-term price movements, interest rates or production and trade flows to follow a cyclical movement lasting about half a century. While Kondratiev was by no means the originator of the long cycle theory, his name is generally associated with those long-term cycles today. J. van Duijn, *The Long Wave in Economic Life* London, Allen and Unwin, 1983 and A. Kleinknecht, *Innovations Patterns in Crisis and Prosperity: Schumpeter's Long Cycle Reconsidered*, Amsterdam, Vrije Universiteit Amsterdam, 1984, contain the best overviews of the early long wave debate.
22. It is worth pointing out that the study of and evidence for long-term movements is a relatively well-established research field within economic history analysis. Economic historians generally distinguish between the pre-industrial era (pre-1790) and the industrial era (post-1780). The latter is primarily related to technological changes and accelerations in the production of industrial goods, whereas in the former, short-term and long-term movements are in the first instance related to meteorological and climatological factors affecting agricultural production. The fundamental difference between the two lies in the fact that in the pre-industrial era one is not confronted with overproduction of industrial goods, but underproduction of agricultural goods. Interestingly, the evidence in relation to these pre-industrial long-term cycles seem rather conclusive and the relationship with meteorological factors significant. The time span of these pre-industrial long-term cycles (approximately thirty years) appears, however, to be totally different from the industrial long waves.
23. As a parenthesis, it is worth noting that this is probably a unique feature of long wave research. In economics, for example, there are, if anything, too few theories to explain the variety and contradictory nature of economic data and the day-to-day working of the 'real', rather than 'economic theory' world. As the 'pure theory' economist wonders: what's wrong with reality?
24. J. Delbeke, 'Recent Long Wave Theories: A Critical Survey', *Futures*, **13**, No. 4 (1981), pp. 246–58, also published in C. Freeman (ed.) *Long Waves in the World Economy*, London, Butterworth, 1983.
25. E. Lederer, *Technischer Fortschritt und Arbeitslosigkeit*, Tübingen, J. C. B. Mohr, 1938, also translated as *Technical Progress and Unemployment*, Geneva, International Labour Office, 1938.
26. A. van der Zwan, 'An Assessment of the Kondratiev Cycle and Related Issues', in S. K. Kuipers and G. J. Lanjouw (eds), *Prospects of Economic Growth*, Amsterdam, North-Holland Publishing Co., 1983.
27. van Duijn, op. cit., note 21.

28. Freeman, op. cit., note 9.
29. Freeman, op. cit., note 15.
30. Freeman, op. cit., note 15.
31. Freeman, op. cit., note 15.
32. Y. Varga, *Two Systems: Socialist Economy and Capitalist Economy*, New York, Internationalist Publisher, 1939; and *Politico-Economic Problems of Capitalism*, Moscow, Progress Publishers, 1966, Russian edition, 1963.
33. Freeman, op. cit., note 15.
34. Freeman, *et al.*, op. cit., note 8, pp. 67–8.
35. Freeman, op. cit., note 9.
36. F. Chesnais, 'Schumpeterian Recovery and the Schumpeterian Perspective—Some Unsettled Issues and Alternative Interpretations' in H. Giersch (ed.), *Emerging Technology: Consequences for Economic Growth, Structural Change and Employment in Advanced Open Economies*, Tübingen, J. C. B.Mohr, 1982.
37. C. Freeman, J. A. Clark and L. L. G.Soete, 'Long Waves, Inventions and Innovations', *Futures*, **13**, No. 4 (1981), pp. 308–22, also published in C. Freeman (ed.), op. cit., note 24.
38. Freeman, op. cit., note 9, p. 14.
39. Writing in 1977, Freeman put it as follows:'there can be little doubt that the 1973 OPEC crisis has led to a vast increase in the effort directed to energy-saving technical change, both in the private and public sectors. Yet the immediate effects of these public pressures and economic incentives were very slight. The main results of the current explosion of energy R & D will probably not be felt until the 1980s, and the same would probably apply to any pressure to swing the system on to a path of relatively greater factor-saving innovation of a specific type, except in extreme cases, such as blockade or war. The argument which is being advanced here is that the 'natural trajectory' of technology is in part supply-determined and may be responsive to major changes from the demand side only with considerable time-lags and imperfections'. Freeman, op. cit., note 15, p. 7.
40. C. Freeman, op. cit., note 15.
41. J. S.Metcalfe, 'Impulse and Diffusion in the Study of Technical Change', *Futures*, **13**, No. 5 (1981), pp. 347–59, also published in C. Freeman (ed.), op. cit., note 24.
42. Metcalfe, op. cit., note 41, p. 21.
43. In particular, the contributions from S. Kuznets, *Secular Movements in Production and Prices*, Boston, Houghton Miflin, 1930, and A. F. Burns, *Production Trends in the United States Since 1870*, New York, NBER, 1934, reprinted 1950.
44. Metcalfe, op. cit., note 41, pp. 102–3.
45. See in particular G. Mensch, op. cit., note 13.
46. Freeman *et al.*, op. cit., note 8, pp. 65–6.
47. Ibid., pp. 66–7.
48. See e.g. F. Chesnais, op. cit., note 36 and A. Kleinknecht, 'Prosperity, Crises and Innovation Patterns', *Cambridge Journal of Economics*, **8**, No. 3, (1984), pp. 308–45; A. Kleinknecht, 'Observations on the Schumpeterian Swarming of Innovations', *Futures*, **13**, No. 4 (1981), pp. 293–307, also published in C. Freeman (ed.), op. cit., 24.
49. In response to Mensch's comments (G. Mensch, 'Long Waves and Techno-

logical Developments in the 20th Century: Comment' in D. Petzina and G. van Roon (eds), *Konjunktur, Krise, Gesellschaft: Wirtschaftliche Wechsellagen und Soziale Entwicklung im 19 und 20 Jahrhundert*, Stuttgart, Klett-Cotta, 1981) on Freeman's Bochum contribution (C. Freeman, J. A. Clark and L. L. G. Soete, 'Long Waves and Technological Developments in the 20th Century' in ibid., pp. 132–69). Freeman lists the twenty-one innovations out of a total of sixty-two which Mensch had omitted from the original list compiled by J. Jewkes *et al.*, *The Sources of Invention*, London, Macmillan, 2nd rev. edn, 1968.

50. As Dosi acknowledges, it was actually Freeman who first, in 'The Determinants of Innovation: Market Demand, Technology and the Response to Social Problems', *Futures*, **11**, No. 3 (1979), pp. 206–15, drew the analogy between the dynamics of technology and scientific procedures and suggested the concept of paradigm change.
51. G. Dosi, 'Technological Paradigms and Technological Trajectories', *Research Policy*, **11**, No. 3 (1983), pp. 147–64; G. Dosi, 'Technological Paradigms and Technological Trajectories: The Determinants and Directions of Technical Change and the Transformation of the Economy' in C. Freeman (ed.), op. cit., note 24.
52. See R. Nelson and S. Winter, *An Evolutionary Theory of Technical Change*, Cambridge, Cambridge University Press, 1982.
53. In fairness to Dosi, it should be pointed out that a set of very similar arguments were also presented in his paper to the Royal College of Art Conference on Long Waves. Dosi's contribution was all the more remarkable because it did attempt to link the concepts of 'paradigm change' and institutional 'match' with some of the work of the French 'regulation' school. He did, however, at least not explicitly, espouse the 'long wave' idea: 'Technology and Conditions of Macro Economic Development: Some Notes on Adjustment Mechanisms and Discontinuities in the Transformation of Capitalist Economies' in C. Freeman (ed.), *Design, Innovation and Long Cycles in Economic Development*, London, RCA, Design Policy Studies, 1984.
54. C. Perez, 'Structural Change and the Assimilation of New Technologies in the Economic and Social System', *Futures*, **15**, No. 4, October (1983), pp. 357–75, also published in C. Freeman (ed.), ibid.
55. Perez, op. cit., note 54, p. 359.
56. C. Freeman, 'Prometheus Unbound', *Futures*, **16**, No. 5 (1984), p. 499.
57. Many references were being made to Freeman's previous writings in the debate about the so-called 'collapse of work'.
58. C. Perez, 'Micro-Electronics, Long Waves and World Structural Change: New Perspectives in Developing Countries', *World Development*, **13**, No. 3 (1985), pp. 441–63.
59. C. Freeman and C. Perez, 'Long Waves and New Technology', *Nordisk Tidsrift for Politisk Ekonomi*, **17** (1984), pp. 5–15.
60. See Freeman's Chap. 3 in C. Freeman and L. Soete, *Information Technology and Employment: An Assessment*, Brussels, IBM, 1985.
61. Freeman and Perez, op. cit., note 59.
62. C. Freeman, 'Keynes or Kondratiev—How Can We Get Back to Full Employment? in P. K. Marstrand (ed.), *New Technology and the Future of Work and Skills*, London, Frances Pinter, 1984, pp. 103–23.
63. See C. Freeman and L. L. G. Soete, 'Managing the Labour Implications of New Technologies', *Employment Growth in the Context of Structural Change*, Paris, OECD, 1984; and C. Freeman and L. Soete, op. cit., note 60.

64. D. Gordon, R. Edwards and M. Reich, *Segmented Work, Divided Workers: The Historical Transformation of Labour in the United States*, Cambridge, Cambridge University Press, 1982; D. Gordon, T. Weisskopf and S. Bowles 'Long Swings and the Nonreproductive Cycle', *American Economic Review*, Papers and Proceedings, **73**, No. 2 (1983), pp. 258–94.
65. M. Aglietta, *A Theory of Capitalist Regulation*, London, New Left Books, 1979; A. Lipietz, 'Conflits de Répartition et Changements Techniques dans la Théorie Marxiste', *Économie Appliquée* (1980); R. Boyer and J. Mistral, *Accumulation, Inflation et Crises*, Paris, Presse Universitaire de France, 1978.
66. G. Dosi, 'Technology and Conditions of Macro Economic Development: Some Notes on Adjustment Mechanisms and Discontinuities in the Transformation of Capitalist Economies' in C. Freeman (ed.), op. cit., note 53.

The publications of Christopher Freeman

With C. T. Saunders and R. W. Eveley, *Industrial Research in Manufacturing Industry, 1959–60*, London, FBI, 1961.

With L. Dicks-Mireaux, C. O'Herlihy, R. Major and F. Blackaby, 'Prospects for the British Car Industry', *National Institute Economic Review*, No. **17**, London, NIESR, 1961, pp. 15–47.

Proposed Standard Practice for Surveys of Research and Development: The Measurement of Scientific and Technical Activities, Paris, OECD, Directorate for Scientific Affairs, 1962.

'Research and Development: A Comparison between British and American Industry', *National Institute Economic Review*, No. 20 (May 1962), London, NIESR, pp. 21–39.

With R. Poignant and I. Svennilson, *Science, Economic Growth and Government Policy*, background paper for Ministerial Meeting on Science, Agenda Item III, Paris, OECD, 1963.

With J. Fuller and A. Young, 'The Plastics Industry: A Comparative Study of Research and Innovation', *National Institute Economic Review*, No. 26 (November 1963), London, NIESR, pp. 22–49.

With A. Young, *The Research and Development Effort in Western Europe, North America and the Soviet Union: An Experimental International Comparison of Research Expenditure and Manpower in 1962*, Paris, OECD, 1965.

'Research and Development in Electronic Capital Goods', *National Institute Economic Review*, No. 34 (November 1965), London, NIESR, pp. 40–91.

With H. Brooks, L. Gunn, J. Saint-Geours and J. Spacy, *Government and Allocation of Resources to Science*, Ministerial Meeting on Science, Paris, OECD, 1966.

'Research, Technical Change and Manpower Forecasting' in B. C. Roberts and H. H. Smith (eds), *Manpower Policy and Employment Trends*, London, Bell,1966.

'Wissenschaftspolitik in Grossbritannien;, *Atomzeitalter* (August 1966), **8**, pp. 233–42; French version: 'Les Options Scientifiques de la Grand Bretagne', *Atomes*, **238** (December 1966), pp. 678–83.

'Foreword' to the collection of papers on *Innovation and the Balance of Payments: The Experience in the Pharmaceutical Industry*, London, Office of Health Economics, 1967.

'Science and Economy at the National Level', paper presented at the Experimental Working Session on Science Policy of the OECD, Jouy-en-Josas, February 1967.

'A Comment on "Expenditures on Research and Development in Poland, 1961 to 1965" by Michael Borowy', *Minerva*, **V**, No. 3 (Spring 1967), pp. 371–5.

'Research Comparisons: Some Limitations of International Comparisons of Research and Development Expenditure', *Science*, **158** (October 1967), pp. 463–8.

With J. H. Killip and R. C. Curnow, 'The British Space Programme: A Reappraisal', paper prepared for National Industrial Space Committee, 1968.

'Science and Economy at the National Level', *Problems of Science Policy*, Paris, OECD, 1968.

With G. Oldham and E. Türkcan, 'The Transfer of Technology to Developing Countries with Special Reference to Licensing and Know-How Agreements', *Proceedings of UNCTAD Second World Conference*, New Delhi, 1 February 1968.

With A. Robertson, R. Curnow, P. Whittaker and J. Fuller, 'Chemical Process Plant: Innovation and the World Market', *National Institute Economic Review*, No. 45 (August 1968), London, NIESR, pp. 29–57.

With A. B. Robertson, 'Innovation and the World Market for New Chemical Plant', *Chemistry and Industry*, 12 October 1968, pp. 1386–91.

The Measurement of Output of Research and Experimental Development: A Review Paper, Paris, UNESCO, 1969.

The Measurement of Scientific and Technological Activities, Paris, UNESCO, 1969.

'National Science Policy', *Physics Bulletin*, **20** (1969), pp. 265–70.

'Science Policy, Technical Progress and Economic Growth' in UN Economic Commission for Europe, *Policies and Means of Promoting Technical Progress* (1969), pp. 1–16.

'Size and Firm, R & D and Innovation', and supplementary paper 'Innovation and Size of Firm', *Proceedings of Conference on Monopolies, Mergers and Restrictive Practices*, Cambridge, King's College, 1969.

'Chemical Process Plant: Innovation and the World Market', *New Technology*, **29** (June 1969).

'Irish Science Policy', paper delivered at Dublin Conference, June 1969, National Science Council of Ireland, 1969.

With C. M. Cooper, O. Gish, C. H. G. Oldham, S. C. Hill, H. Singer and R. C. Desai, 'Draft Introductory Statement for the World Plan of Action for the Application of Science and Technology to Development', Annexe II of *Science and Technology for Development: Proposals for the Second UN Development Decade*, UN Department of Economic and Social Affairs, New York, UN, 1970.

'Innovation and Coupling Systems in Research and Development', in Maurice Goldsmith (ed.), *Technological Innovation and the Economy*, New York, John Wiley, 1970, Chap. 18, pp. 177–87.

'Technology Assessment and its Social Context', *Studium Generale*, **24** (1971), pp. 1038–50.

'Industrial Innovation', *IVA*, **42**, No. 7, Stockholm, 1971, p. 235.

With B. G. Achilladelis and A. B. Robertson, 'A Study on Innovation in the Chemical Industry: A Preliminary Report', *Chemistry and Industry* (February 1971), **10**, pp. 269–73.

'Hamlet without the Prince of Denmark', *Patent Office Examining Staff Magazine*, May 1971, pp. 17–21.

'Industrial Innovation: the Key to Success?', *Electronics and Power*, **42** (August 1971), pp. 297–301.

'The Role of Small Firms in Innovation in the United Kingdom Since 1945', Committee of Inquiry on Small Firms, Research Report No. 6, HMSO, October 1971.

With C. H. G. Oldham, C. M. Cooper, T. C. Sinclair and B. G. Achilladelis, 'The Goals of R & D in the 1970s', *Science Studies*, **1**, No. 3 (October 1971), pp. 357–406. Also Italian version: 'Gli Objettivi della Ricerca e dello Sviluppo negli anni

1970–1980', in *Rapporto sulla Scienza*, Etas Kompass, Italy, February 1972, pp. 51–121.

With A. B. Robertson, P. Jervis, B. G. Achilladelis, A. Horsely, R. C. Curnow and J. K. Fuller, 'Success and Failure in Industrial Innovation', Report on Project SAPPHO, London, Centre for the Study of Industrial Innovation, 1972.

'European Science Policy', paper prepared for Council of Europe's 3rd Parliamentary and Scientific Conference, Lausanne, 11–14 April 1972.

'A Study of Success and Failure in Industrial Innovation' in B. R. Williams (ed.), *Science and Technology in Economic Growth*, Proceedings of Conference held by the International Economic Association, St. Anton, Austria, London, Macmillan, 1973, pp. 227–45.

'The International Science Race' in D. O. Edge and J. N. Wolfe (eds), *Meaning and Control: Essays in Social Aspects of Science and Technology*, Tavistock Publications, 1973.

With H. S. D. Cole, M. Jahoda, K. L. R. Pavitt (eds), *Thinking about the Future: A Critique of 'The Limits to Growth'*, London and Brighton, Chatto and Windus/ Sussex University Press, May 1973. Also published in *Futures*, March and April 1973. American edition: *Models of Doom: a Critique of 'The Limits to Growth'*, New York, Universe Books, 1973. German edition: *Zukunft aus dem Computer? Eine Antwort auf 'Die Grenzen des Wachstums'*, Luchterhand, 1973. French edition: *L'Anti-Malthus: une Critique de 'Halte à la Croissance'*, Paris, Seuil, 1974. Also author of Chap. 1, 'Introduction: Malthus with a Computer', and co-author of Chap. 6, 'The Capital and Industrial Output Subsystem'.

'Inter-Governmental Cooperation and the Future', *In Europe Now: Cooperation in Research and Technology*, Proceedings of 1973 Symposium of the R & D Society, September 1973, pp. 24–30.

'Malthus with a Computer', *World Issues*, **25** (Winter 19734), pp. 6–8.

The Economics of Industrial Innovation Harmondsworth, Penguin Modern Economic Texts, 1974. Spanish edition: *La Teoria Economica de la Innovacion Industrial*, Madrid, Penguin Alianza, 1975. Second edition: London, Frances Pinter, 1982.

With R. Rothwell, A. Horsley, V. T. P. Jervis, A. B. Robertson and J. F. Townsend, 'SAPPHO Updated, Project SAPPHO Phase II', *Research Policy*, **3**,No. 3 (1974), pp. 258–91.

'Malthus sur Ordinateur', *La Recherche*, **43** (March 1974), pp. 251–6.

With P. A.Julien, 'Malthus in, Malthus out?', *L'Actualité Économique*, **2** (April–June 1974), pp. 232–44.

'Science Policy, Systems Analysis and World Models', *UK Regional Systems Bulletin* (Summer 1974), pp. 16–17.

'The Luxury of Despair: a Reply to Robert Heilbroner's Human Prospect', *Futures*, **6**, No. 6 December 1974), pp. 450–62; also published in R. Jones (ed.), *Readings from Futures 1974–80*, Guildford, Westbury House, 1981, pp. 41–53.

With I. D. Miles and M. Jahoda, *Progress and Problems in Social Forecasting*, SSRC, 1976.

With B. V. A. Roling, A. M. Weinberg and H. F. York, *Technological Innovation: A Socio-political Problem*, Proceedings of the Symposium 'Control of Technological Development', organized on the occasion of the 15th Anniversary of the Twente University of Technology, 29–30 November, 1976, Boerderijcahier 7701, Twente, 1977.

With K. L. R. Pavitt, 'The Current International Economic Climate and Policies for Technical Innovation', report prepared for the Six Countries Programme on Government Politics towards Technological Innovation in Industry, TNO, The Netherlands, 1977.

'The Kondratiev Long Waves, Technical Change and Unemployment', *Structural Determinants of Employment*, **2**, Paris, OECD, 1977, pp. 181–96.

'Economics of Research and Development' in E. Spiegel-Rosing and D. de Solla Price (eds), *Science Policy Studies in Perspective*, London, Sage Publications, 1977, Chap. 7, pp. 223–75.

'Technical Change and Unemployment', paper presented to Conference on 'Science, Technology and Public Policy: An International Perspective', University of New South Wales, 1–2 December 1977, published in Conference Proceedings.

With M. Jahoda (eds), *World Futures: The Great Debate*, Martin Robertson, 1978.

'Preface to *Policies for the Stimulation of Industrial Innovation: Vol. 1, Analytical Report*, Paris, OECD, 1978, pp. 5–14.

Innovation and Size of Firm, Occasional Paper No. 1, Science Policy Research Centre, School of Science, Griffith University, Australia, 1978.

With R. C. Curnow, 'Product and Process Change arising from the Microprocessor Revolution and Some of the Economic and Social Consequences', Keynote Address to Institution of Mechanical Engineers, May 1978.

'Government Policies for Industrial Innovation', J. D. Bernal Memorial Lecture, Birkbeck College, May 1978.

With R. C. Curnow, 'Technical Change and Employment: A Review of Postwar Research', paper prepared for Manpower Services Commission, June 1978.

'Obstacles in the Responsiveness of Science and Technology to Problems of Society', paper presented at ESIST Seminar, organized by EEC, Compiègne, France, 19–20 October 1978.

'Technology and Employment: Long Waves in Technical Change and Economic Development', Holst Memorial Lecture, Eindhoven, The Netherlands, December 1978.

'Technical Innovation and British Trade Performance' in F. Blackaby (ed), *Deindustrialisation*, Heinemann/NIESR, Economic Policy Papers 2 (1979), pp. 56–77.

'Technical Change and Unemployment' in S. Encel and J. Ronayne (eds), *Science, Technology and Public Policy: An International Perspective*, Rushcutters Bay, Oxford, Pergamon Press, 1979, pp. 53–76.

'Microelectronics and Unemployment' in *Automation and Unemployment*, papers presented at ANZAAS Symposium, Sydney, The Law Book Co. Ltd., 1979, pp. 99–113.

'The Kondratiev Long Waves, Technical Change and Unemployment' in *Structural Determinants of Employment and Unemployment, II*, reports prepared for Experts Meeting in Paris, 7–11 March 1977, Paris, OECD, 1979, pp. 181–96.

'The Determinants of Innovation: Market Demand, Technology and the Response to Social Problems', *Futures*, **11**, No. 3, 1979, pp. 206–15.

'Social and Economic Impact of Microelectronics', paper presented at Workshop on Technology Assessment: its Role in National and Corporate Planning', Sydney, Australia, 25–26 July 1979, published in proceedings, Canberra, Australian Government Publishing Service, 1979, pp. 13–24.

'Unemployment and Technical Change', Thomas McLaughlin Memorial Lecture, Dublin, November 1979.

With U. Colombo, R. Nelson, K. Pavitt and N. Rosenberg, *Technical Change and Economic Policy: Science and Technology in the New Economic and Social Context*, Paris, OECD, 1980.

'Unemployment and Government' in T. Forester (ed.), *The Microelectronics Revolution: the Complete Guide to the New Technology and its Impact on Society*, Oxford, Basil Blackwell, 1980, pp. 308–17.

'The Diffusion of Electronic Technology in the Economic System', paper prepared for 60th Anniversary of Netherlands Electronics and Radio Society, *Journal of the NERS*, **3** (1980), pp. 169–76.

'Microelectronics, Competition and Employment', paper prepared for Conference in Karlsruhe, February 1980, published as: 'Die Mikroelektronik: Wettbewerb und Arbeitsmarke', ISI Jahreskolloquiums zum Thema 'Mikroelektronik, Wettbewerb und Beschaftigung—eine Bilanz'.

'Government Policy' in K. L. R. Pavitt (ed.), *Technical Innovation and British Economic Performance*, London, Macmillan, 1980, pp. 310–25.

With A. F. Bonfiglioli and R. W. Cahn, 'Policy for Energy and Materials Conservation in Relation to the Aluminium Industry', final report to Science Research Council, mimeo. May 1980.

'British Trade Performance and Technical Innovation', *Technical Innovation and National Economic Performance*, papers from Workshop held at Institute of Production, Aalborg University Centre, 8 December 1980, Denmark, Aalborg University Centre, 1981, pp. 1–25.

With J. A. Clark and L. L. G. Soete, 'Long Waves and Technological Developments in the 20th Century' in D. Petzina and G. van Roon (eds), *Konjunktur, Krise, Gesellschaft: Wirtschaftliche Wechsellagen und Sociale Entwicklung im 19 und 20 Jahrhundert*, Stuttgart, Klett-Cotta, 1981, pp. 132–69.

'Introduction to the New Technology', *Managing Technology in the '80s*, Part 2—Complete Proceedings, London, LAMSAC, 1981, pp. 12–22.

'Technology and Employment: Long Waves in Technical Change and Economic Development' in H. S. D. Cole *et al.* (eds), *Methods for Development Planning: Scenarios, Models and Micro Studies*, Paris, UNESCO, 1981, pp. 121–34.

'Some Economic Implications of Microelectronics', *Technology and Employment: the Impact of Microelectronics*, papers from Workshop held at Institute of Production, Aalborg University Center, Denmark, 19 February 1981, pp. 7–54.

'Policies for Technical Innovation in the New Economic Context' and 'Concluding Comments' in P. H. Kristensen and R. Stankiewicz (eds), *Technology Policy and Industrial Development in Scandinavia*, proceedings of Workshop held in May 1981, Research Policy Institute, Lund University, Sweden, and Institute of Economics and Planning, Roskilde University Center, Denmark, pp. 21–44 and 209–17.

With J. A. Clark and L. L. G. Soete, 'Long Waves, Inventions and Innovations', *Futures*, **13**, No. 4, 1981, pp. 308–22.

(Ed.), 'Technical Innovation and Long Waves in World Economic Development', *Futures*, **13**, No. 4 (August 1981) and No. 5 (October 1981), special issues.

'Débat: les Technologies Nouvelles, sont-elles a l'Origine de la Crise Économique', *La Recherche*, **12**, No. 125 (September 1981), pp. 982–9. Also published as 'Debat: Son las Neuvas Tecnologias un Elemento Generador de la Crisis Economica', *Mundo Cientifico*, **8** (November 1981), pp. 921–5.

'Innovation as an Engine of Economic Growth: Retrospect and Prospects' in H. Giersch (ed.), *Emerging Technologies: Consequences of Economic Growth,*

Structural Change and Employment, proceedings of Kiel Symposium, 1981, Tübingen, J. C. B. Mohr (Paul Siebeck), 1981, pp. 1–32.

'The Economic Implications of Microelectronics' in C. D. Cohen (ed.) *Agenda for Britain 1: Micro Policy Choices for the 80s*, Oxford, Phillip Allan, 1982, pp. 53–88.

With J. A. Clark and L. L. G. Soete, *Unemployment and Technical Innovation: A Study of Long Waves and Economic Development*, London, Frances Pinter, 1982. Spanish edition: *Desempleo e Innovacion Tecnologica: un estudio de las ondas largas y el desarrollo economico*, Madrid, Ministerio de Trabajo y Seguridad Social, 1985.

'Science, Technology and Unemployment', lecture at Imperial College of Science and Technology, 9 February 1982, published as Paper No. 1 in the series Papers in Science, Technology and Public Policy, Imperial College/SPRU, 1982.

With L. L. G. Soete and J. F.Townsend, 'Fluctuations in the Numbers of Product and Process Innovation 1920–1980', paper presented at second Workshop on Patent and Innovation Statistics, Paris, OECD, 28–30 June 1982.

'Recent Developments in Science and Technology Indicators: A Review', report to International Development Research Centre, Ottawa, November 1982.

(Ed.), *Long Waves in the World Economy*, enlarged edition of two special issues of *Futures* (August and October 1981) containing additional papers by European economists, London, Butterworth, 1983; second edition, London, Frances Pinter, 1984.

With J. F. Townsend and V. M. Walsh, 'The Determinants of Technical Change in the Chemical Industry: Demand-Pull or Technology-Push?' in Stephen F. Frowen (ed.), *Controlling Industrial Economies*, Vienna Institute for Comparative Studies (VICES), London, Macmillan, 1983, pp. 83–108.

'Technological Change and the New Economic Context' in S. Hill and R. Johnston (eds), *Future Tense: Technology in Australia*, University of Queensland Press, 1983, pp. 47–67.

With L. L. G. Soete, 'Cambio Tecnologico y Politicas de Ajuste', *Papeles de Economia Espanola*, special edition on 'Politicas para una Recuperacion Prolongada', Madrid, Raycar, S.A. Matilde Hernandez, 1983, pp. 386–95.

'Design and British Economic Performance', paper presented at the Design Centre, Department of Design Research, the Royal College of Art, Haymarket, London, 13–15 April 1983.

'Changes in the External Environment of Industrial R and D', paper presented at European Industrial Research Management Association Annual Conference, Interlaken, Switzerland, 18–20 May 1983, published in EIRMA Conference Papers Vol. XXVIII, *The Role of Industrial R and D in the 1980s*, Paris, EIRMA, 1983, pp. 35–44.

'Long Waves and Technical Innovation', paper presented at the TNO Conference on 'Technology and Economic Development', 15 September 1983, *Conference Proceedings*, Netherlands, TNO, 1983.

'Technical Innovation, Economic Growth and Employment', paper presented at the International Symposium on 'Analysis of the Crisis', Majorca, 13–15 October 1983.

'Keynes or Kondratiev: How can we get back to Full Employment?' paper prepared for British Association for the Advancement of Science, Annual Meeting, 22–26 August 1983, published in P. K. Marstrand (ed.), *New Technology and the Future of Work and Skills*, London, Frances Pinter, 1984, pp. 103–23.

'Inovavao de ciclos longos de desenvolvimento economico', *Ensios FEE*, **5**, No. 1 (1984), pp. 5–20.

'New Technology, Structural Change and International Competition', paper presented to meeting of Swiss members of Club of Rome, La Tour-de-Peilz, 23–24 January 1984.

With L. L. G. Soete, 'Managing the Labour Implications of New Technologies', paper prepared for OECD Conference on Employment Growth in the Context of Structural Change, Paris, OECD, February 1984.

'Capital Shortage, Technology and Unemployment', paper prepared for the Seminar on Technical Change and Employment, Aalborg Institute of Production, 27–28 April 1984.

(Ed.), *Design, Innovation and Long Cycles in Economic Development*, Design Research Publications No. 1, London, Royal College of Art, 1984.

With C. Perez, 'Long Waves and New Technology', *Nordisk Tidsskrift for Politisk Ekonomi*, **17** NOPEK, July 1984, pp. 5–15.

'Prometheus Unbound', *Futures*, **16**, No. 5 (October 1984), pp. 495–507. *Die Computer revolution in den langen Zyklen der okonomischen Entwicklung*, München, Carl Friedrich von Siemens Stiftung, 1984.

'The Computer Revolution: Nightmare or Utopia', Whidden Lectures, McMaster University, Canada, October 1984, to be published.

'Some Recent Developments in Science and Technology Indicators: A Summary Review', paper prepared for the Stifterverband für die Deutsche Wissenschaft (June 1984).

'Future Employment Prospects in Print and the Communications Industries', paper prepared for the cancelled IPEX International Conference, National Exhibition Centre, Birmingham, 4–12 September 1984.

'The Role of Technical Change in National Economic Development' in J. B. Goddard (ed.) *Technological Change, Industrial Restructuring and Regional Development*, Allen and Unwin, 1985, pp. 100–15.

Schumpeter und Kondratieff: Die Computerrevolution und Lange Zykler der Ükonomischen Entwicklung, München, Carl Friedrich von Siemens Stiftung, 1985.

'Long Waves of Economic Development' in T. Forrester (ed.), *The Information Technology Revolution*, London, Basil Blackwell, 1985, pp. 602–16.

(Ed.), *Technological Trends and Employment: 4. Engineering and Vehicles*, London, Gower, 1985.

With L. L. G. Soete, *Information Technology and Employment: An Assessment*, Brussels, IBM, 1985.

'Information Technology as a Change of the Techno-Economic Paradigm', paper presented to International Hightech Forum, Basle, 28–29 November 1985.

'Theories of the Long Wave', *Long Waves, Depression and Innovation*, Laxenburg, IIASA, 1985.

With L. L. G. Soete, 'New Technologies, Investment and Employment Growth' in *Employment Growth and Structural Change* Paris, OECD, 1985.

With L. L. G. Soete, 'Theories of The Long Wave', *Long Waves, Depression, and Innovation: Implications for National and Regional Economic Policy*, Proceedings of the Sienna/Florence Meeting, 26–30 October 1983, International Institute for Applied Systems Analysis A-2361, Laxenburg, March 1985.

'Biotechnology and the Diffusion of New Products, Processes and Systems', paper prepared for OECD Workshop on Long-Term Economic Impacts of Biotechnology, Rome, 1–2 April 1985.

'The Economics of Innovation', *IEE Proceedings*, **132**, pt. A, No. 4 (July 1985).
'Economic Theory, Technical Change and Quality of Life', Kikawada Foundation Memorial Lecture, September 1985.
'Technical Change and Unemployment', paper prepared for International Symposium on Microelectronics and Labour, Tokyo, 17 September 1985.
'Technical Innovation and the New Economic Context', paper prepared for International Symposium on Microelectronics, Tokyo, 17 September 1985.
With L. L. G. Soete (eds), *Technical Change and Full Employment*, London, Basil Blackwell, 1986, forthcoming.
'The Science Policy Research Unit', to be published in R. J. Blin-Stoyle (ed.), *University of Sussex Silver Jubilee Book*, 1986.
'Innovation', 'Long Swings in Economic Growth' and 'Structural Unemployment', Contributions to *The New Palgrave*, London, Macmillan, 1986, forthcoming.

Notes on contributors

Asa Briggs, the distinguished historian, is Provost of Worcester College, Oxford. Among the founding faculty in the University of Sussex, he was its Vice-Chancellor from 1967 to 1976, and was Chairman of the Supervisory Committee of the Science Policy Research Unit (SPRU) from its inception. In 1976 he was made a Life Peer.

Norman Clark received his Ph.D. in Economics from the University of Edinburgh. Between 1968 and 1972, he was a Research Fellow at SPRU, before becoming Lecturer in Development Economics at the University of Glasgow. From 1977 to 1978, he was Director of the Technology Planning and Development Unit at the University of Ife, Nigeria. In 1980, he was appointed Reader and Chairman of the History and Social Studies of Science Group at the University of Sussex, and moved to the chairmanship of SPRU's postgraduate programme in 1982. He is the author of a book and a number of articles in the field of science policy analysis.

Charles Falk received his Ph.D. in Physics from Carnegie-Mellon University, and subsequently worked at the Brookhaven National Laboratory, where he became Associate Director. Following many years in the National Science Foundation, he has recently retired from the Directorship of the Science Resources Studies Division. He was a Visiting Fellow at SPRU between 1972 and 1973. He is a specialist in the collection and analysis of science and technology resource data, and in the development of science and technology indicators.

Amilcar O. Herrera is a graduate of the University of Buenos Aires and the Colorado School of Mines, where he took degrees in geology and economic geology. He has been a professor at the University of La Plata, the University of Buenos Aires, the University of Chile, and at the Fundacion Bariloche. He is at present Director of the Institute of Geosciences and Science Policy in the University of Campinas, Brazil.

Marie Jahoda received her doctorate in psychology from the University of Vienna. She has taught at the University of New York, Brunel University and at the University of Sussex, from which she holds the title of Professor Emeritus. She is the author of several books and many articles in social psychology.

Roy MacLeod is Professor of History at the University of Sydney. He received his degrees from Harvard and Cambridge, and became in 1966 one of the first Research Fellows appointed to SPRU. In 1970, he became foundation Reader in the History and Social Studies of Science at the University of Sussex. Between 1978 and 1982, he was foundation Professor of Science Education at the Institute of Education, University of London. He has written and edited several books and articles in the social history of science and technology, and has taught for many years in the social studies of science.

Richard Nelson received his Ph.D. in Economics from Yale University. He has served on the staff of the Rand Corporation and the (US) Council of Economic Advisers, and has participated in many study groups concerned with science and technology policy. He has taught at Oberlin College and at Carnegie-Mellon University, and has been on the staff of Yale since 1968. He is now Professor of Economics, and Director of the Institution for Social and Policy Studies.

Keith Pavitt read engineering and industrial management at Cambridge, and economics and public policy at Harvard. He was a staff member in the Directorate for Scientific Affairs at the Organisation for Economic Cooperation and Development (OECD), and a Visiting Lecturer at Princeton, before becoming a Senior Fellow at SPRU, in 1971. In 1984, he succeeded Christopher Freeman as Reginald M. Philips Professor of Science Policy. He is now Deputy-Director of SPRU.

George F. Ray is a Senior Research Fellow at the National Institute of Economic and Social Research (London), which he joined in 1957. He has been a member of the Editorial Board of the *National Institute Economic Review* since 1968, and a visiting professor (in energy economics) at the University of Surrey since 1975. He was President of the Association d'Institute Européens de Conjoncture Économique between 1978 and 1984, and Editor of The *Business Economist* from 1969 to 1974. He has written extensively on energy commodities, innovation, industrial economics and forecasting, and the diffusion of innovations.

Nathan Rosenberg has been Professor of Economics at Stanford University since 1974. He was educated at Rutgers University, the University of Wisconsin, and Oxford University. He has written extensively on the economics of technological change. His recent books include *Perspectives on Technology* (1976) and *Inside the Black Box* (1982).

Jean-Jacques Salomon is Professor of Technology and Society, and Director of the Centre Science, Technologie et Société at the Conservatoire National des Arts et Métiers, Paris. He graduated from the Sorbonne, where he did his doctorate in philosophy and history of science. He founded and was for two decades Head of the Science and Technology Policy Division at OECD. He has been a visiting professor in several universities, including MIT, Harvard and Montreal, and a Visiting Fellow of Clare Hall, Cambridge. His books include *Science and Politics* (1973), *The Research System* (1972–4), and *Prométhée empêtré* (1982).

Luc Soete is Reader in European Science and Technology Policy at SPRU. He obtained his degrees in economics at the University of Ghent (1972) and the University of Sussex (1978). Before joining SPRU in 1980, he held positions at the University of Antwerp and the Institute of Development Studies at the University of Sussex. During 1984–5, he was a visiting associate professor at Stanford University. Over the last five years, he has been one of the closest collaborators of Christopher Freeman, with whom he has written a number of books and papers.

Bruce Williams was the first Director of the Technical Change Centre in London, from its foundation in 1981 to 1986, and is a Visiting Professor at the Imperial College of Science and Technology. He graduated from the University of Melbourne, and later served as Lecturer in Economics at the University of Adelaide, and Senior Lecturer in Economics at Queen's University, Belfast. He became Professor of Economics at the University College of North Staffordshire, 1949–59, and at the University of Manchester, 1959–67. He was Vice-Chancellor and Principal of the University of Sydney from 1967–81, and was knighted in 1980. He has written several books concerned with economics, education and technical change.

Index